Exploitation of Unconventional Oil and Gas Resources - Hydraulic Fracturing and Other Recovery and Assessment Techniques

Edited by Kenneth Imo-Imo Eshiet

Published in London, United Kingdom

IntechOpen

Supporting open minds since 2005

Exploitation of Unconventional Oil and Gas Resources - Hydraulic Fracturing and Other Recovery and Assessment Techniques
http://dx.doi.org/10.5772/intechopen.74901
Edited by Kenneth Imo-Imo Eshiet

Contributors
Keshawa Shukla, Shehzad Ahmed, Alvinda Hanamertani, Muhmmad Rehan Hashmet, Freddy Humberto Escobar, Tanja Srebotnjak, Maman Hermana, Chow Weng Sum, Deva Prasad Ghosh, Ahmed Ouenes, Yamina Aimene, Chad Hammerquist, Mohit Paryani, Aissa Bachir, Kenneth Imo-Imo Eshiet

First published in London, United Kingdom, 2019 by IntechOpen
IntechOpen is the global imprint of INTECHOPEN LIMITED, registered in England and Wales, registration number: 11086078, The Shard, 25th floor, 32 London Bridge Street
London, SE19SG - United Kingdom
Printed in Croatia

British Library Cataloguing-in-Publication Data
A catalogue record for this book is available from the British Library

Additional hard and PDF copies can be obtained from orders@intechopen.com

Exploitation of Unconventional Oil and Gas Resources - Hydraulic Fracturing and Other Recovery and Assessment Techniques
Edited by Kenneth Imo-Imo Eshiet
p. cm.
Print ISBN 978-1-83881-107-5
Online ISBN 978-1-83881-108-2
eBook (PDF) ISBN 978-1-83881-109-9

For EU product safety concerns:
IN TECH d.o.o., Prolaz Marije Krucifikse Kozulić 3, 51000 Rijeka, Croatia, info@intechopen.com or visit our website at intechopen.com.

Meet the editor

Dr. Kenneth Imo-Imo Israel Eshiet is Senior Consultant at SEEED (Sustainable Energy Environmental and Educational Development) with research links with the University of Leeds and University of Wolverhampton, UK. He is an expert with broad experience in geomechanical stimulation of reservoirs and subsurface flow. He is a Chartered Engineer in the UK and holds a doctorate degree in Civil Engineering (geotechnics) from the University of Leeds, in tandem with other postgraduate qualifications in management and education. His current interest lies in the development and application of numerical/analytical tools to civil, structural, fluid dynamics, and geotechnical engineering problems. He is Editor of the American Journal of Civil Engineering and a reviewer of several journals, including Computers and Geotechnics, Environmental Earth Sciences, and Bulletin of Engineering Geology and the Environment. He resides in West Yorkshire, UK.

Contents

Preface

Unconventional hydrocarbon production has drawn more attention than ever recorded in history. The reason is simply the promise of a vast amount of reserve, the potential for cleaner fuels, and advances in technology, which now enable drilling and extraction of hydrocarbons under challenging conditions. Evolution in technologies has also made it possible to exploit unconventional formations more efficiently and economically. Generally, unconventional extraction of oil and gas can only be achieved if the formation is adequately stimulated to encourage outward flow of hydrocarbon fluids. This is necessary because of the formation rock's ultra-low permeability and porosity as well as other uncharacteristic rock and reservoir fluid properties. Sundry methods have been successfully applied in stimulating unconventional reservoirs. Hydraulic fracturing still remains one of the major ways of accomplishing this; the principles adopted in this method, which is very much reflected in most of the other stimulation approaches, is fundamentally the introduction of fluids with special properties into reservoirs. Fluids are injected under different states—e.g., high or low pressure, high or low velocity (injection rate), high or normal temperature, and acidic or non-acidic.

An additional way to enhance reservoir productivity is through the drilling of directional wells. Directional (i.e., inclined and horizontal wells) are considered better than vertical wells in terms of several aspects of reservoir performance. In comparison to vertical wells, the productivity of a reservoir is greatly improved where hydraulic fracturing or other stimulation approaches based on the same concept are implemented through directional wells.

This book sheds light on selected themes that are crucial to the stimulation of unconventional hydrocarbon reservoirs. These are treated under seven areas:

- reservoir simulation strategies;
- fracturing fluids and fluid systems;
- diffusion and mixing of fracturing and hydrocarbon fluids;
- assessment of reservoir performance and fracture characterization;
- integrating surface drilling data in stimulation designs;
- estimating brittleness and hydrocarbon content: correlating brittleness and fracture density and detecting the existence and quantity of hydrocarbons; and
- health and safety.

The introductory chapter, "Developments in the Exploitation of Unconventional Hydrocarbon Reservoirs," is a conspectus of key aspects of unconventional hydrocarbon production, and up-to-date developments and challenges based on the above-named thematic areas. The following are covered: different types of reservoir stimulation approaches; fracturing and reservoir fluids/fluid systems; drivers of fluid behavior with respect to their multiphasic and multicomponent mixing and transport; well test analysis; and environmental impact and health and safety.

Chapter 2, "CO_2 Foam as an Improved Fracturing Fluid System for Unconventional Reservoirs," describes the characteristics and applications of CO_2 foam-based fracturing fluids. Foam-based fluids have a number of advantages over other types of fracturing fluids. Among other benefits, CO_2 foam fracturing fluids are suitable for water-sensitive formations, facilitate water flowback/recovery, reduce the quantity of produced water, and improve proppant placement and distribution. The rheology of CO_2 foam is explained, supported by corresponding models and physical experimental studies that describe the foam fluid behavior.

Chapter 3, "Thermodynamics of Thermal Diffusion Factors in Hydrocarbon Mixtures," presents a thermodynamic model of thermal diffusion factors for hydrocarbon mixtures, which relies on the linear transport of intermolecular forces while considering the effects of energy and molecular mass and size. In furtherance, this model is used to examine the behavior of binary hydrocarbon fluid mixtures.

Chapter 4, "Well Test Analysis for Hydraulically Fractured Wells," illustrates how the *Tiab's direct synthesis* (TDS) technique (a well test analysis method) is applied in interpreting pressure tests and assessing the performance of hydraulically fractured reservoirs and their associated wells. The TDS technique can be used for fracture characterization and provides reliable estimates of fracture parameters, for instance, fracture conductivity and half-length.

Chapter 5, "Surface Drilling Data for Constrained Hydraulic Fracturing and Fast Reservoir Simulation of Unconventional Wells," presents a workflow demonstrating the use of surface drilling data in constructing reservoir models. These models are applied in designing hydraulic fracturing operations and for unconventional reservoir simulation. The main surface drilling data considered are rate of penetration (rop), torque (t) and weight-on-bit (WOB).

Chapter 6, "Elastic-based Brittleness Estimation from Seismic Inversion," introduces procedures for determining rock brittleness, fracture density, and hydrocarbon bearing in fractured reservoirs. In the first approach, brittleness is indicated by the parameter *brittleness average* (BA). BA can be used to identify zones of high fracture density, also referred to as the brittle area. It is calculated from elastic properties—Poisson's ratio and Young's modulus—which are in turn derived from seismic data through seismic inversion. Poisson's ratio is determined from P-wave (V_P) and S-wave (V_S) velocities, while Young's modulus is expressed in terms of bulk modulus and Poisson's ratio. In the second approach, rock brittleness, fracture density, and the nature of the formation lithology is indicated by the parameter *scaled inverse quality factor of P-wave* (SQp). A related parameter, *scaled inverse quality factor of S-wave* (SQs), is used as an indicator of the presence of hydrocarbons.

Chapter 7, "Human Health Risks of Unconventional Oil and Gas Development using Hydraulic Fracturing," looks at the potential human exposures to emissions (and attendant health and safety risks) as a result of hydraulic fracturing operations in unconventional oil and gas reservoirs. This is viewed from the perspective of both occupational and public health and safety.

Kenneth Imo-Imo Israel Eshiet PhD, CEng, MICE, FHEA
Sustainable Energy Environmental and Educational Development (SEEED) Ltd,
Sugarland, Texas, USA

Section 1

Introduction

Chapter 1

Introductory Chapter: Developments in the Exploitation of Unconventional Hydrocarbon Reservoirs

Kenneth Imo-Imo Eshiet

1. Introduction

Hydrocarbon reservoirs contain fossil fuels and constitute a major proportion of sources of energy worldwide. In the past, extraction of oil and gas was mainly restricted to conventional reservoirs which underlie a sealing caprock or rock formation with lower permeability and consist of rock and fluid with characteristics that readily allow the flow of oil and gas into wellbores. These reservoirs are easily assessed and contain sufficient pressure such that the external and additional drive necessary to push the hydrocarbon fluids to the surface are not exigent. Conventional reservoirs are recognised by their structural layout, stratification and rock and fluid properties. Typically, they comprise three major parts: a cap rock, a source rock and a reservoir rock. The cap rock is the impermeable rock layer that seals the boundaries of the top and sides and entraps the hydrocarbons within the reservoir. Hydrocarbons are formed in the source rock (normally limestones or shales) which contains kerogen, an insoluble and solid organic matter. The reservoir rock is the permeable and porous layer containing hydrocarbon fluids generated in the source rock. Over a protracted period, oil and gas formed in source rocks migrate to reservoir rocks, a process that is essential for the existence and validity of reservoir rocks.

With the advent in advanced technology and increasing need for more and cleaner energy, oil and gas production has been extended to unconventional reservoirs. Generally, unconventional reservoirs are difficult to produce. They are mainly composed of very tight source rocks containing hydrocarbons that have not migrated to reservoir rocks. These ultra-tight source rocks are termed unconventional reservoirs. Fundamentally, unconventional and conventional reservoirs are differentiated based on the migration of hydrocarbons from source rocks. Conventional reservoirs are rock formations that are recipient of hydrocarbons from source rocks, while unconventional reservoirs are source rocks containing hydrocarbons that cannot be naturally released to reservoir rocks. Nonetheless, the term unconventional reservoirs broadly cover reservoir rocks which are problematic to produce, for instance, tight reservoir rocks (tight sandstones, tight limestone, etc.) and heavy oil reservoir rocks.

Artificial lift is a standard method of instigating flow from the reservoir through the wellbore. This technique decreases the bottomhole pressure (BHP) while increasing the pressure in the reservoir, thereby raising the rate of well production. It is inevitably required at a certain time during the life of an oil/gas field due

Themes
i. Reservoir stimulation strategies
ii. Fracturing fluids and fluid systems
iii. Diffusion and mixing of fracturing and hydrocarbon fluids
iv. Well test analysis
v. Health, safety and the environment

Table 1.
Outline of some fundamental aspects of unconventional oil and gas production.

to diminishing flow rates or for the removal of liquids to enable gas flow. Often, artificial lifts are sufficient for conventional reservoirs as a means of actuating or boosting flow; however, when applied in isolation, the same effects are not obtained in unconventional reservoirs. A vast amount of hydrocarbons are trapped within unconventional reservoirs. These reservoirs possess tremendous economic potential which can only be realised if they are properly stimulated. A range of reservoir stimulation methods are now available which have rendered unconventional reservoirs commercially viable. Some of these methods (e.g. hydraulic fracturing) can be applied to a broad spectrum of reservoirs, whereas others (e.g. some forms of acidisation) have limited applications. Since the boom in production from unconventional reservoirs, great strides in development have been made, with an increasing number of source rocks and depleted reservoir rocks subject to being produced. This has also raised concerns in relation to the impact on atmospheric, aquatic, land and underground environments; climate change; economic viability; technology requirements; health and safety; and sustainability. Ongoing studies are geared towards improving the process of exploiting unconventional reservoirs and increasing value for money while ensuring minimal levels of pollution and contamination to the environment, as well as risk to humans, and flora and fauna. The scope of areas considered in terms of the exploitation of unconventional resources is apparently inexhaustible, especially when viewed from a microscopic perspective. However, these can be harmonised into a more condensed list of themes. Some key aspects regarding the exploration and production of unconventional reservoirs are discussed in this chapter. These are encompassed within the subjects of discourse itemised in **Table 1**.

2. Reservoir stimulation strategies

Reservoir stimulation is simply described as the induction of formations to improve hydrocarbon production. This is can be accomplished by repairing the formation damage, especially at the vicinity of the wellbore, and/or changing the natural state of the rock or fluids to increase reservoir productivity. The advent in oil and gas production from unconventional reserves has given rise to the development of several stimulation approaches. This inventory (of methods) has expanded over the years, with considerable improvements made to boost the effectiveness and efficiency of a sizeable number of them. Some techniques are focused on repairing damages that have impaired the conductivity of rocks surrounding the wellbore, and some artificially create additional channels to enable easy flow of reservoir fluids towards the wellbore, while others alter the properties of reservoir fluids to make them less adhesive to host rocks and to encourage nonviscous-like fluid flow into wellbores. Numerous stimulation techniques are currently employed in practice. These include hydraulic fracturing, surfactant flooding and treatment, water imbibition, acidisation, thermal stimulation and treatment and electrokinetics potential.

2.1 Hydraulic fracturing

Hydraulic fracturing is the artificial initiation and proliferation of fractures by high-pressured injection of fluid into the rock. The fracturing fluid is pumped into the rock at a pressure that surpasses the rock failure stress [1]. The operation is aimed at generating new fractures, reopening/expanding and extending the reach of existing fractures and increasing their connectivity. The concept of this well stimulation technique was foremostly introduced by Hubbert and Willis [2] and has since been developed into an effective and widely used method of increasing oil and gas reservoir productivity. Hydraulic fracturing has been successfully applied in various conventional (for enhanced oil recovery/gas recovery (EOR/EGR)) and unconventional reservoirs/source rocks, such as oil shales and tight rocks (i.e. tight oil/gas sandstones, limestones, shale, etc.) (e.g. [1, 3, 4]). Various designs of hydraulic fracturing operations are currently implemented in practice; an example is multicluster-multistage horizontal well fracturing (e.g. [5, 6]).

2.2 Surfactant treatment/flooding

Surfactant flooding is a technique used to lower the amount of oil entrapment in the rock matrix. Surfactants are amphiphilic organic chemicals comprising hydrophobic and hydrophilic compounds. When injected into ultra-low permeability hydrocarbons rocks, they reduce the oil viscosity, pore capillary forces and the interfacial tension between water and oil and decrease oil-wet wettability [7–9].

2.3 Thermal stimulation/treatment

Thermal stimulation and treatment are particularly useful for reservoirs with high-viscosity fluids and for the release of methane-rich gases from gas hydrates. Heavy oil formations contain high-density and high-viscosity fluids, which make them even more challenging to produce. Viscosity is heat dependent, having an inverse relationship with temperature. This implies that when there is a rise in temperature, the heavy reservoir oils become less viscous and, thus, more mobile. Thermal stimulation methods for heavy oils include steam flooding, steam-assisted gravity drainage (SAGD) and cyclic steam stimulation [10, 11]. Methane-rich gases naturally trapped in gas hydrates can be produced by raising the temperature of the formation. In practice, steam or hot brine is injected to heat the deposit; however, other innovative methods, e.g. microwave heating and electromagnetic heating, may be adopted [12].

2.4 Acidisation

This is the injection of acids into a reservoir to dissolve the rock matrix. Dissolution of portions of the rock creates wormholes while increasing its permeability and porosity [13]. Acid fracturing and matrix acidisation are the two main acidisation methods adopted as stimulation techniques [14]. The most commonly used types of acid are hydrochloric acid (HCL) and hydrofluoric acid (HF). These are often not applied singly; rather they are combined together or with other types of acids (e.g. organic acids) (e.g. [15, 16]). Carbonate rocks such as limestone, dolomites and carbonate-rich shales are readily soluble in HCL. On the contrary, silicate or quartz-based rocks (e.g. sandstones) are not soluble in HCL; they react favourably with HF.

2.5 Water imbibition

This is the process of absorption of water as the wetting phase into rock, in which saturation by the wetting phase rises while that by the non-wetting phase reduces. This phenomenon can be induced by water flooding [17] or surfactant flooding [18, 19]. A shift in wettability from oil-wet to water-wet increases water imbibition. Water imbibition into porous rock improves oil recovery by displacing trapped hydrocarbons at both the rock surface and the pores. An example of this is *spontaneous imbibition* which is used successfully for oil recovery in shale reservoirs [20, 21].

2.6 Electrokinetics potential

Passing direct current (DC) through an oil reservoir generates an electrokinetics potential which causes electrophoresis, electromigration, electrochemical reaction, electro-osmosis and Joule heating [22]. The collective actions of these mechanisms improve formation rock permeability and porosity through the dislodgement and removal of pore linings in the form of colloids, thereby increasing pore sizes and creating new flow paths [22].

3. Fracturing fluids and fluid systems

Fracturing fluids are used for hydraulic fracturing and are injected into formations either as highly pressurised fluids or as acid-based fluids used to etch the walls of existing or newly formed fractures, creating additional flow channels. There are four main categories of fracturing fluids: water-based, oil-based, acid-based and foam-based fracturing fluids. The characteristics of fracturing fluids affect the pattern of fractures formed. Viscosity and density are the major properties that primarily determine the fluid behaviour. When designing a fracturing fluid system, it is imperative that, at least, the following are taken into consideration: fluid viscosity, fluid rheology, rock conductivity, cost, the impact on the environment, proppant carrying capacity, friction loss, compatibility of the fluid with the formation rock and the net pressure drop in the fractures.

The first type of fluids preferred for hydraulic fracturing was oil-based, including hydrocarbons such as gasoline and kerosene [23, 24]. Oil-based fracturing fluids are low in viscosity and generally need to be mixed with chemicals for its quality to be improved. They are excellent fracturing fluid alternatives for water-sensitive formations. Oil-based fluids are shown to be recyclable and compatible with drilling fluids and can be fully recovered during clean-up [25].

Water-based fluids are the most predominant fracturing fluids and in many ways better alternatives to oil-based fluids. The advent of water-based fracturing fluids introduced the petroleum industry to safer and cheaper substitutes to oil-based fluids. They can be classified as slickwater, cross-linked fluids, uncross-linked (linear) fluids and viscoelastic surfactant (VES) fluids [24]. Water-based fracturing fluids are aqueous, consisting mostly of water mixed with varying proportions of chemical additives and proppants [26]. The added chemicals may serve as viscosifiers or friction reducers. Acid-based fluids are suitable for formation rocks that are acid-soluble.

Acid-based fracturing fluids frequently used in practice are hydrochloric acid (HCL), hydrofluoric acid (HF) and organic acids. Carbonate rocks (e.g. limestone, dolostone and carbonate-rich shale) and silicate-rich rocks (e.g. sandstone) are soluble in HCL and HF, respectively [13, 27]. Most formation rocks are not exclusively one or the other; therefore, in many instances, an acid blend (mud acid) comprising a combination of more than one type of acid is used.

Foam-based fluids are composed of a mixture of gas and liquid phases with a very high percentage of the gas fraction within the range of $52\% \leqq F^g \leqq 96\%$, where F^g is the percentage composition of gas [28]. Nitrogen (N_2) and carbon dioxide (CO_2) are the common gas phases used in practice, while either acids or water or polymer or alcohol (methanol) constitute the liquid phase [29]. Foam-based fluids are also appropriate for water-sensitive formations and have been successfully applied in shale gas reservoirs [29, 30]. The proppant carrying capacity of foam-based fluids greatly exceeds (by$\approx$ 85%) that of water-based fluids. Their application requires a considerably less amount of water, and there is less liquid to recover at the end of the fracturing operation. Moreover, foam is recyclable and reusable, implying a reduction in waste and cost [29, 30]. The demerits are mainly the high initial costs and logistic requirement and the decrease in viscosity in high temperatures [30].

4. Diffusion and mixing of fracturing and hydrocarbon reservoir fluids

Hydrocarbon reservoirs are often multicomponent and multiphase. This means that in their natural state, there are variations in composition of reservoir fluids, occurring longitudinally and/or vertically. Key drivers of changes in reservoir

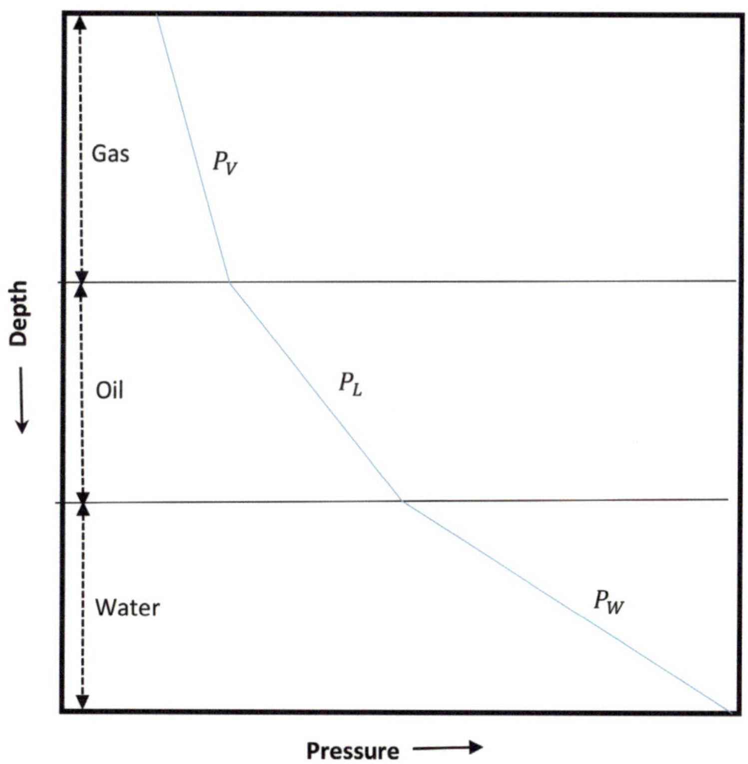

Figure 1.
Pressure gradient of multicomponent reservoir fluids without capillary effect. P V, P L and P W are the vapour, liquid and water pressure, respectively [32].

fluid composition are gravity, capillary forces and temperature gradients [31]. Gravity and capillary effects are major factors influencing variations in composition with depth ([32, 33]). Due to gravity, reservoir liquid hydrocarbons lie atop aquifers, which is a reflection of the differences in density between the two fluids (hydrocarbons are less dense and immiscible in water). In terms of hydrocarbons, the gas phase lies above the liquid phase, and their individual pressure gradient is dependent on their corresponding densities [32]. An idealised form of this, ignoring capillary effects, is shown in **Figure 1**.

Discounting capillary action renders the illustration in **Figure 1** unrealistic for formation rocks which are porous and therefore composed of pore spaces. Capillary forces due to surface tension within the pores act in opposition to external forces such as gravity. In addition, reservoirs are likely to contain a mixture of multi-component fluids at the different phases (gas and liquid), such as the occurrence of pockets of heavy hydrocarbons or injected fluids within the predominant fluid type. This will change the composition with depth in any or both of the following ways: firstly, capillary forces prompt the occurrence of a transition region, which is an overlap consisting of two or more phases instead of the sudden change shown in **Figure 1** and, secondly, the compositional gradient of the reservoir fluid is altered because of the changes in its components. A modified pressure gradient profile which also accounts for capillarity and compositional variation is given in **Figure 2**.

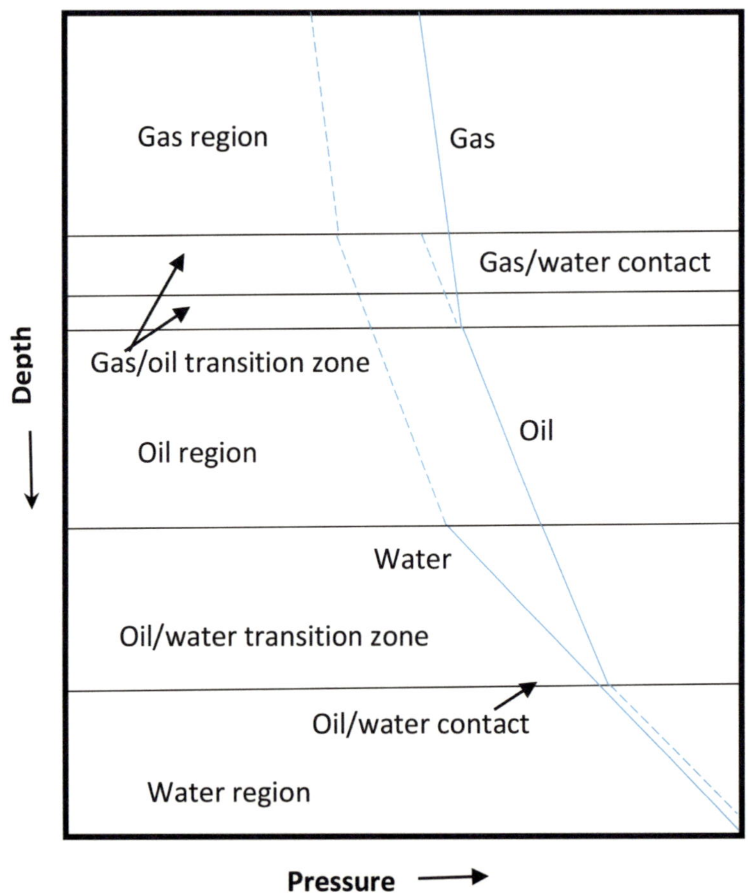

Figure 2.
Pressure gradient of multicomponent reservoir fluid with the combined effects of gravity, capillarity and variation in composition [32].

Temperature gradients in formations introduce an extra dimension to the behaviour of reservoir fluids. The effect of variations in temperature induces convection and thermal diffusion. For small temperature gradients collinear with the gravity vector, convection can be neglected [31]. Thermal diffusion, also known as 'Soret effect', is the separation of a non-convective mixture due to a thermal gradient [34]. In other words, there is movement of material during the occurrence of thermal gradients resulting in corresponding concentration gradients of the constituents of the fluid mixture. This process is measured by the thermal diffusion coefficient, α. Thermal diffusion can have a substantial impact on variations in composition of reservoir fluids [34, 35]; it may increase or attenuate compositional variation vertically and horizontally.

The mixing of injected fracturing fluids with in situ and/or other fracturing fluids affects the constituent composition and variation of reservoir hydrocarbon fluids and the stimulation process during enhanced oil/gas recovery. The introduction of alien fluid(s) into the formation sets off a mixing mechanism and displacement of resident fluids, which improve hydrocarbon production. Controlling factors include, but not limited to, the injected/resident fluid properties (e.g. rheology, density and viscosity), formation rock properties, the reservoir condition (thermal gradient, pre-existing fluid compositional variation) and other drivers such as capillarity and diffusion. The practicability of this process involving a three-phase fluid system (scCO_2-brine-oil) is demonstrated in Jiménez-Martínez et al. [36], where supercritical CO_2 (as an injection fluid) is used to restimulate an oil-wet shale formation containing brine and hydrocarbon as the major resident fluids.

5. Well test analysis

Well test analysis—also known as pressure transient test analysis—consists of methods of finding and evaluating information regarding the well and reservoir. More specifically, it involves the manipulation and measurement of flow rates and pressures which can then be linked to well and reservoir conditions. The process primarily entails altering the well flow while monitoring temporal variations in pressure [37] or vice versa [38]. The magnitude and changes in pressure are used to deduce the reservoir size, wellbore damage, boundaries and heterogeneities (e.g. fault positions), reservoir pressure at the drainage region, well deliverability, flow rate [37] and other reservoirs or related parameters such as hydraulic connectivity, skin effect and permeability.

Well test analysis is the process of assessment and interpretation of data obtained from well tests using a variety of techniques. A diagnostic set of plots consisting of trends of pressure and its derivative (relating to time) against time is a common tool that facilitates the interpretation of well tests (pressure transient tests) [37]. The trend of pressure on these plots can then be used to determine the flow regime. For instance, flow regime *specialised* plots (ΔP *vs* $f(\Delta t)$) aid the identification of flow regimes [39] (e.g. radial, linear, bilinear and spherical flows), where ΔP is the change in pressure and $f(\Delta t)$ is a flow regime-specific function which is dependent on changes in time. An alternative approach (the Homer method) introduced by Homer [40] to overcome certain shortcomings of *specialised* plots measures ΔP against a superposition time specific to a given flow regime.

Over the years, the manner of conducting well test analysis has evolved. Types of well test analysis methods (interpretation methods) include straight line, pressure type curve and pressure derivative analysis and deconvolution; these are listed in order of the period they were developed. Detailed description of these methods is given in Gringarten [39].

6. Health, safety and the environment

Undoubtedly, the production of unconventional hydrocarbons is attendant with several benefits. These include an increase in the global quantity of energy sources, cheaper prices, the accessibility to relatively cleaner energies, etc. Despite these merits, there are various drawbacks subsumed under environmental impacts and health and safety. Occupational and public health risks are thriving in the oil and gas industry. While there are risk exposures common to hydrocarbon production in general, there are concerns specific to unconventional reservoirs. This can be viewed from three standpoints: the environmental impact, health and safety. Environmental effects deal with negative changes inflicted on the surroundings and far-reach zones (sub-surface, surface and atmospheric regions) as a result of production activities. Health and safety issues focus on the effect on humans and are divided into two facets: occupational and public.

6.1 Occupational health and safety

Workers are customary exposed to numerous hazards. Transport-related activities are reported as the highest cause of accidents due to the prolific movement of people, equipment, chemicals, hydrocarbon produce, etc. [41]. There is also a risk of explosions from inflammable and high-pressured fluids and contact with (by inhaling) hazardous constituent compounds of hydrocarbons such as hydrogen sulphide and crystalline silica usually used as proppants for hydraulic fracturing [41, 42].

6.2 Public health

A myriad of studies are available that support the narrative linking of unconventional hydrocarbon production with a range of human health problems. Examples of these effects are cancer, mental stress, eye irritation, respiratory disease, cardiovascular disease and congenital defects [43, 44]. Whereas the potential for these diseases is undoubted, evidence-based and scientifically proven cause-and-effect relationships between unconventional production activities and community health are lacking. The constraints, in some cases, are the inaccessibility to reliable data or the biased interpretation of data or the use of non-validated protocols to generate and analyse data. It is suggested that credible studies should be based on standard epidemiological procedures [43], which properly identifies stressors and their sources, the pathways through which humans are contacted and the health impact. Potential exposures include air, soil, surface water and groundwater contamination; odours; noise; seismic events and earthquakes; increase in traffic and accident rate; and water shortage [44].

6.3 Environmental effects

The impact of unconventional hydrocarbon production on the environment is principally focused on adverse alterations in the ecosystem, surrounding water bodies (e.g. aquifers and surface waters) and land and air contamination/pollution (such as greenhouse gas emissions). These can be categorised as air, land, water, biodiversity (ecosystems and wildlife) and waste impacts [45]. Air impacts involve emissions of volatile chemicals and greenhouse gases which reduce air quality, and waste impacts deal with challenges associated with the management of wastes produced by unconventional hydrocarbon production. Human exposures are facilitated through contact with affected media (i.e. air, land, water, biodiversity and waste).

7. Summary comments

Stimulation of unconventional hydrocarbon reservoirs to enable or improve production is inexorable. There are a plurality of reasons for this; the primary ones are encapsulated in the constraints that hinder the access of the target reservoir and/or source rock and the peculiarity of both formation rocks and fluids. The distinctive nature of unconventional formations are manifested through, for instance, rocks with ultra-low permeability and porosity, the presence of heavy oils as constituent reservoir fluids and the multiphase and multicomponent composition of the formation. Stimulation approaches used in practice are wide ranging. Some of these—e.g. hydraulic fracturing—are age long and have evolved into well-developed methods, whereas others, e.g. spontaneous imbibition, are advancing at a fast pace.

Hydraulic fracturing is traditionally used to artificially create additional flow channels by injecting fluids at high pressures; however, aspects of these techniques are adopted or used in tandem with other stimulation methods. Acid fracturing, for example, is one of the two major acidisation techniques and involves the injection of acids at pressures high enough to generate fractures while dissolving and etching their surfaces. The central objectives of each stimulation method and its limitation are determinants of the choice of fracturing fluid or fluid system. Obviously, it is expected that acid-based fluids would be used for acid fracturing operations. Likewise, either foam-based or oil-based fracturing fluids are superior options for water-sensitive formations. The behaviour of in situ reservoir fluids including their interactions with injected fluids (in terms of mixing, diffusion, etc.) influences the effectiveness of the recovery process and the recyclability and reusability of the introduced fluids. Pivotal drivers of reservoir fluid behaviour include the properties of the injected/resident fluid and formation rock, the reservoir condition, gravity, capillarity and diffusion.

Other important aspects regarding the exploitation of unconventional hydrocarbon formations are health, safety and the environmental effect. This is generally considered in terms of occupational and public health and safety and the environmental impact of drilling and production activities. Studies on occupational health and safety are fairly established; there seem to be sufficient evidence to substantiate correlations linking health and safety hazards with incidences of accident in the industry. Also, standardised environmental impact assessments have made it possible to identify and measure changes in surrounding and far-reach areas through, for instance, the use of indicators. Conversely, there are several grey areas with respect to threats to public health and safety, since the validity of many investigative studies is disputable because they are apparently subjective, incredible and therefore inconclusive.

Author details

Kenneth Imo-Imo Eshiet
Sustainable Energy Environmental and Educational Development (SEEED) Ltd, Sugarland, Texas, USA

*Address all correspondence to: kenieshiet@yahoo.com

References

[1] Bennour Z, Ishida T, Nagaya Y, Chen Y, Nara Y, Chen Q, et al. Crack extension in hydraulic fracturing of shale cores using viscous oil, water, and liquid carbon dioxide. Rock Mechanics and Rock Engineering. 2015;**48**:1463-1473. DOI: 10.1007/s00603-015-0774-2

[2] Hubbert KM, Willis DG. Mechanics of hydraulic fracturing. PetroleumTransactions, AIME. 1957;**210**:153-166

[3] Gandossi L. An overview of hydraulic fracturing and other formation stimulation technologies for shale gas production. In: Policy Report by the Joint Research Centre of the European Commission. Report EUR 26347 EN; ISBN: 978-92-79-34729-0 (pdf), ISSN: 1831-9424 (online). 2013. DOI: 10.2790/99937

[4] Reinicke A, Rybacki E, Stanchits S, Huenges E, Dresen G. Hydraulic fracturing stimulation techniques and formation damage mechanisms—Implications from laboratory testing of tight sandstone–proppant systems. Chemie der Erde. 2010;**70**(S3):107-117. DOI: 10.1016/j.chemer.2010.05.016

[5] Lin R, Ren L, Zhao J, Wu L, Li Y. Cluster spacing optimization of multi-stage fracturing in horizontal shale gas wells based on stimulated reservoir volume evaluation. Arabian Journal of Geosciences. 2017;**10**:38. DOI: 10.1007/s12517-016-2823-x

[6] Sookprasong PA, Hurt RS, Gill CC. Downhole monitoring of multicluster, multistage horizontal well fracturing with fiber optic distributed acoustic sensing (DAS) and distributed temperature sensing (DTS). In: The International Petroleum Technology Conference; 10-12 December 2014; Kuala Lumpur, Malaysia. Paper IPTC-17972-MS. 2014. DOI: 10.2523/IPTC-17972-MS

[7] Abdulelah H, Mahmood SM, Al-Mutarreb A. Effect of anionic surfactant on wettability of shale and its implication on gas adsorption/desorption behavior. Energy and Fuels. 2018;**32**(2):1423-1432. DOI: 10.1021/acs.energyfuels.7b03476

[8] Kamal MS, Hussein IA, Sultan AS. Review on surfactant flooding: Phase behavior, retention, IFT, and field applications. Energy and Fuels. 2017;**31**(8):7710-7720. DOI: 10.1021/acs.energyfuels.7b00353

[9] Mirchi V, Saraji S, Goual L, Piri M. Dynamic interfacial tensions and contact angles of surfactant-in-brine/oil/shale systems: Implications to enhanced oil recovery in shale oil reservoirs. In: SPE Improved Oil Recovery Symposium; 12-16 April 2014; Tulsa, Oklahoma, USA. Paper SPE-169171-MS. 2014

[10] Alajmi HM. Assessment of development methods for a heavy oil sandstone reservoir [PhD thesis]. Imperial College London; 2013

[11] Curtis C, Kopper R, Decoster E, Guzman-Garcia A, Huggins C, Knauer L, et al. Heavy-oil reservoirs. Oilfield Review. 2002;**14**(3):30-51

[12] Cranganu C. In-situ thermal stimulation of gas hydrates. Journal of Petroleum Science and Engineering. 2009;**65**(1-2):76-80. DOI: 10.1016/j.petrol.2008.12.028

[13] Shafiq MU, Mahmud HB. Sandstone matrix acidizing knowledge and future development. Journal of Petroleum Exploration and Production Technology. 2017;7(4):1205-1216. DOI: 10.1007/s13202-017-0314-6

[14] Abdullah KS. Acidizing oil wells, a sister-technology to hydraulic fracturing: Risks, chemicals and regulations [PhD thesis]. University of California; 2016

[15] Alhesan JSA, Amer MW, Mashall M, Jackson WR, Gengenbach T, Qi Y, et al. A comparison of the NaOH-HCl and HCl-HF methods of extracting kerogen from two different marine oil shales. Fuel. 2019;**236**:880-889. DOI: 10.1016/j.fuel.2018.09.058

[16] Hill AD, Sepehrnoori K, Wu PY. Design of the HCL preflush in sandstone acidizing. SPE Production & Facilities. 1994;**9**(2):115-120. DOI: 10.2118/21720-PA

[17] Behrouz H, Mohsen M, Abouzar M-P, Esmaeil H. Experimental study of dynamic imbibition during water flooding of naturally fractured reservoirs. Journal of Petroleum Science and Engineering. 2019;**174**:1-13. DOI: 10.1016/j.petrol.2018.11.008

[18] Miller C, Tong S, Mohanty KK. A chemical blend for stimulating production in oil-shale formations. In: Paper URTeC: 2900955, Presentation at the Unconventional Resources Technology Conference; 23-25 July 2018; Houston, Texas, USA. 2018. DOI: 10.15530/URTEC-2018-2900955

[19] Zeng T, Miller CS, Mohanty K. Application of surfactants in shale chemical EOR at high temperatures. In: Paper Presented at SPE Improved Oil Recovery Conference; 14-18 April 2018; Tulsa, Oklahoma, USA. Paper SPE-190318-MS. Society of Petroleum Engineers; 2018. DOI: 10.2118/190318-MS

[20] Cheng Z, Wang Q, Ning Z, Li M, Lyu C, Huang L, et al. Experimental investigation of countercurrent spontaneous imbibition in tight sandstone using nuclear magnetic resonance. Energy and Fuels. 2018;**32**(6):6507-6517. DOI: 10.1021/acs.energyfuels.8b00394

[21] Wang D, Butler R, Liu H, Ahmed S. Flow-rate behavior and imbibition in shale. SPE Reservoir Evaluation and Engineering. 2011;**14**(4):485-492

[22] Alklih MY, Ghosh B, Al-Shalabi EW. Tight reservoir stimulation for improved water injection—A novel technique. In: Paper Presented at International Petroleum Technology Conference; 19-22 January 2014; Doha, Qatar. Paper IPTC-17536-MS. 2014. DOI: 10.2523/IPTC-17536-MS

[23] Al-Muntasheri GA. A critical review of hydraulic fracturing fluids for moderate- to ultralow-permeability formations over the last decade. SPE Production & Operations. 2014;**29**(4):243-260. DOI: 10.2118/169552-PA

[24] Li L, Al-Muntasheri GA, Liang F. A review of crosslinked fracturing fluids prepared with produced water. Petroleum. 2016;**2**(4):313-323. DOI: 10.1016/j.petlm.2016.10.001

[25] Hlidek BT, Meyer R, Yule K, Joerg W. A case for oil-based fracturing fluids in Canadian Montney unconventional gas development. In: SPE Annual Technical Conference and Exhibition; 8-10 October 2012; San Antonio, Texas, USA. Paper SPE-159952-MS. 2012. DOI: 10.2118/159952-MS

[26] King GE. Hydraulic fracturing 101: What every representative, environmentalist, regulator, reporter, investor, university researcher, neighbor and engineer should know about estimating frac risk and improving frac performance in unconventional gas and oil wells. In: Paper Presented at SPE Hydraulic Fracturing Technology Conference; 6-8 February; The Woodlands, Texas, USA. Paper SPE-152596-MS. Society of Petroleum Engineers; 2012. DOI: 10.2118/152596-MS

[27] Morsy SS, Sheng JJ, Soliman MY. Improving hydraulic fracturing of shale formations by acidizing. In: SPE Eastern Regional Meeting; 20-22 August 2013; Pittsburgh, Pennsylvania, USA. Paper SPE-165688-MS. 2013. DOI: 10.2118/165688-MS

[28] Hutchins RD, Miller MJ. A circulating-foam loop for evaluating foam at conditions of use. SPE Production & Facilities. 2005;**20**(4):286-294. DOI: 10.2118/80242-PA

[29] Gandossi L, Von Estorff U. An overview of hydraulic fracturing and other formation stimulation technologies for shale gas production. In: Policy Report by the Joint Research Centre of the European Commission. Report EUR 26347, ISBN: 978-92-79-53894-0, ISSN: 1831-9424. 2015. DOI: 10.2790/379646

[30] Wanniarachchi WAM, Ranjith PG, Perera MSA. Shale gas fracturing using foam-based fracturing fluid: A review. Environmental Earth Sciences. 2017;**76**:91. DOI: 10.1007/s12665-017-6399-x

[31] Faissat B, Knudsen K, Stenby EH, Montel F. Fundamental statements about thermal diffusion for a multicomponent mixture in a porous medium. Fluid Phase Equilibria. 1994;**100**:209-222. DOI: 10.1016/0378-3812(94)80010-3

[32] Wheaton RJ. Treatment of variations of composition with depth in gas-condensate reservoirs. SPE Reservoir Engineering. 1991;**6**(2):239-244. DOI: 10.2118/18267-PA

[33] Lee S-T. Capillary-gravity equilibria for hydrocarbon fluids in porous media. In: SPE Annual Technical Conference and Exhibition. Paper SPE-19650-MS. San Antonio, Texas; 8-11 October 1989. DOI: 10.2118/SPE-19650-MS

[34] Shukla K, Firoozabadi A. A new model of thermal diffusion coefficients in binary hydrocarbon mixtures. Industrial and Engineering Chemistry Research. 1998;**37**:3331-3342. DOI: 10.1021/ie970896p

[35] Shukla K. Statistical thermodynamics of thermal diffusion factors in binary hydrocarbon mixtures—An application. Molecular Physics. 2017;**115**(9-12):1253-1263. DOI: 10.1080/00268976.2016.1275857

[36] Jiménez-Martínez J, Porter ML, Hyman JD, Carey JW, Viswanathan HS. Mixing in a three-phase system: Enhanced production of oil-wet reservoirs by CO_2 injection. Geophysical Research Letters. 2016;**43**:196-205. DOI: 10.1002/2015GL066787

[37] Fanchi JR. Chapter 12: Well testing. In: Shared Earth Modeling. Oxford: Butterworth-Heinemann, Elsevier; 2002. DOI: 10.1016/B978-075067522-2/50012-4

[38] Ramey HJ Jr. Advances in practical well-test analysis. Journal of Petroleum Technology. 1992;**44**(6):650-659. DOI: 10.2118/20592-PA

[39] Gringarten AC. From straight lines to deconvolution: The evolution of the state of the art in well test analysis. SPE Reservoir Evaluation and Engineering. 2008;**11**(1):41-62. DOI: 10.2118/102079-PA

[40] Horner DR. Pressure build-ups in wells. In: Proc. Third World Petroleum Congress, Section II; 28 May-6 June 1951; The Hague. 1951. pp. 503-523

[41] Floris VV. Management of health and safety hazards in oil and gas. In: SPE Western North American and Rocky Mountain Joint Meeting; 17-18 April 2014; Denver, Colorado. Paper SPE-169509. Society of Petroleum Engineers; 2014. DOI: 10.2118/169509-MS

[42] Esswein EJ, Breitenstein M, Snawder K, Kiefer M, Sieber WK. Occupational exposures to respirable crystalline silica during hydraulic fracturing. Journal of Occupational and Environmental Hygiene. 2013;**10**:347-356. DOI: 10.1080/15459624.2013.788352

[43] Delvin DJ, Beatty P, White RD, Naufal Z, Sarang SS. Collaborative

industry initiative to assess potential health effects related to the production of unconventional resources. In: SPE International Conference on Health, Safety, and Environment; 17-19 March 2014; Long Beach, California, USA. Paper SPE-168518-MS. Society of Petroleum Engineers; 2014. DOI: 10.2118/168518-MS

[44] Health Protection Scotland (HPS). A Health Impact Assessment of Unconventional Oil and Gas in Scotland. Vol. 2. Appendices. Health Protection Scotland; 2016. ISBN: 978-1-873772-50-8

[45] Marsters P, Alvarez FC, Barido DP, Siegner L, Kammen DM. An Analysis of the Environmental Impacts of the Extraction of Shale Gas and Oil in the United States with Applications to Mexico. Berkeley: Energy and Resources Group, University of California. 2015

Section 2

Fracturing and Reservoir Fluid Systems

Chapter 2

CO_2 Foam as an Improved Fracturing Fluid System for Unconventional Reservoir

Shehzad Ahmed, Alvinda Sri Hanamertani and Muhammad Rehan Hashmet

Abstract

Unconventional reservoirs have gained substantial attention due to huge amount of stored reserves which are challenging to produce. Innovative recovery techniques include horizontal drilling coupled with hydraulic fracturing are required to optimize the production of hydrocarbons. There are numerous concerns associated with the utilization of conventional water-based polymeric solutions for fracturing shales. However, the gas utilization has been found as an exceptional stimulation approach providing various benefits. CO_2 foam, an energized fracturing fluid, has been used to overcome the limitation of conventional fracturing fluid. CO_2 foam is able to enhance hydrocarbon production by addressing the critical issues associated with the conventional technique. The rheological property of CO_2 foam fracturing fluid is a key factor controlling the efficiency of overall processes. Different models describing the foam flow behavior have been produced and numerous investigations have been conducted to explain the rheological behavior of foam for fracturing purpose. Various process variables, such as foam quality, temperature, pressure, shear rate, surfactant concentration, and salinity strongly affect foam rheology behavior giving an impact on designing foam fracturing fluid at required fracturing conditions. In-depth analysis and information gathering are substantially required to ascertain the performance of CO_2 foam as an improved fracturing fluid system.

Keywords: fracturing fluid, shales, CO_2 foam, foam rheology

1. Introduction to unconventional reservoirs

Global energy consumption has been increasing rapidly whilst existing oil and gas fields are being depleted day by day. In addition, insufficient amount of hydrocarbons produced from conventional reservoirs to fulfill the increasing energy demand has led to global challenges. Due to these factors and also environmental reasons, the use of natural gas that is considered as a green energy, is demanding. The large volume of natural gas stored in tight formation such as shale and tight sand has been practically developed recently. According to the report of EIA Annual Energy Outlook 2015, the total energy consumption in 2040 will rise to 105.7 quadrillion Btu from 97.1 quadrillion Btu which is about 8.9% of the total energy

IntechOpen

consumption [1]. Most growth is found in the consumption of renewable energy and natural gas. However, energy from the renewable and sustainable resources cannot compete with the nonrenewable and cheap fossil fuel energy in technical and economic aspects. Therefore, unconventional reservoir is considered as an immediate alternative for overcoming the production decline of the conventional reservoirs. These unconventional reservoirs originally come in different forms which include shale gas and oil, tight gas and oil, coal bed methane (CBM), shale oil, and natural gas hydrates as shown in **Figure 1(a)** [2]. The illustration of global gas resources including the general features and the worldwide endowment for each resource is presented in **Figure 1(b)**. Endowment is the sum of undiscovered gas, reserves, and the cumulative gas production. The endowment for the natural gas as shown in **Figure 1(b)** is about 68,000 trillion cubic feet (Tcf) and 70% of it is approximately in shale gas and tight gas reservoir [2].

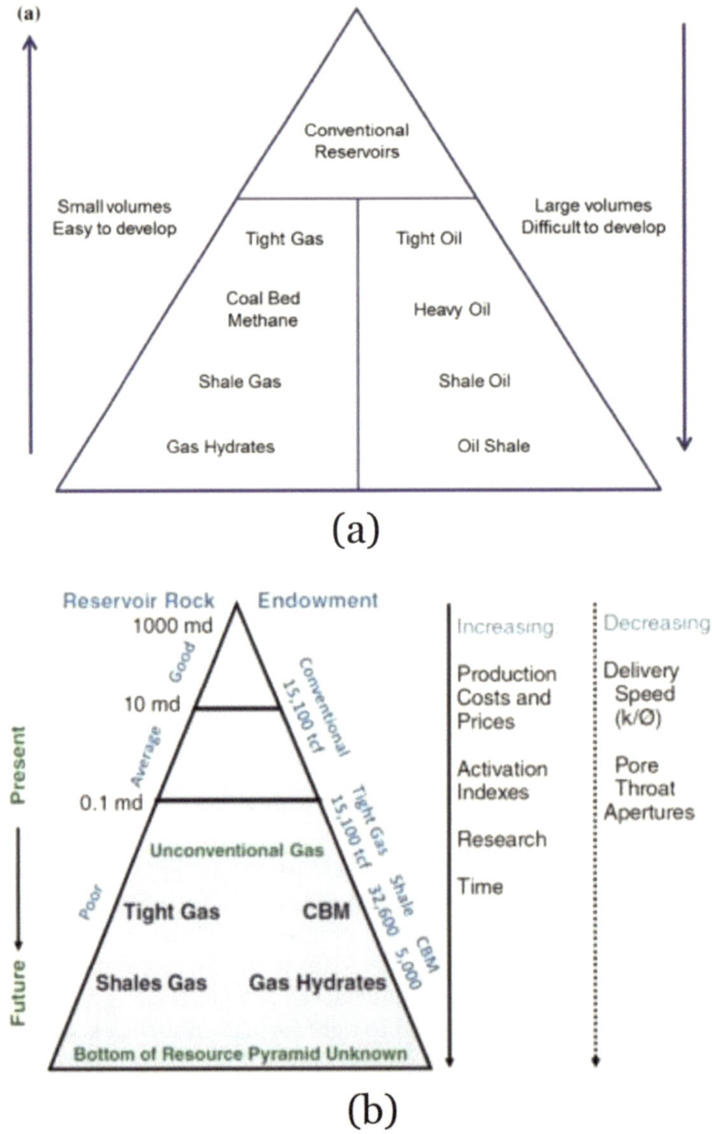

Figure 1.
Pyramid of world oil and gas resources for (a) oil and gas, and (b) natural gas with endowment [2].

Unconventional reservoirs have been significantly important due to their huge amount of stored hydrocarbon. Unconventional reservoirs are defined as those hydrocarbon reservoirs that require stimulation techniques including the alteration of rock permeability or fluid viscosity in order to develop them economically and to produce the hydrocarbon at commercial rates [3]. Unconventional reservoirs, however, come with various challenges that include a complex system having hydraulically induced fractures, natural fractures, and a complex matrix system comprising of different minerals and kerogen [4]. It is of great interest to develop and refine new and existing techniques to recover more oil and gas from these types of unconventional reservoir.

Tight gas is the natural gas present in sandstone or limestone having very low matrix permeability, less than 0.1 millidarcy (mD), and porosity of less than 10% [5]. Shale gas is the trapped natural gas produced from the shale formation with minimal migration. Moreover, coalbed methane is methane gas trapped in coal beds or seams which is stored on the coal internal surface during the coalification process. Natural gas has been called "sweet gas" because of the absence of hydrogen sulfide content which makes it different from the typical conventional gas reservoir [6]. According to the estimate of Energy Information Administration (EIA), recoverable gas in the world is about 7299 trillion cubic feet (Tcf) [7]. The production of natural dry gas produced from shales in the US in 2015 was about 9.96 Tcf which is about 43% of the total US gas production. A continuous increasing trend in shale gas production was observed from 1999 to 2015 as shown in **Figure 2** [8].

The permeability range of shale gas and tight gas reservoir is usually between nano-Darcy (nD) and micro-Darcy (μD). The size of pore throat in shale is in nanometer and there are some cracks present which assist connection between pores [9]. The shale gas reservoir possesses high capillary pressure, high irreducible wetting phase saturation, low porosity, and extremely low permeability [9]. In order to produce gas commercially from these extremely low permeability reservoirs, horizontal drilling and hydraulic fracturing stimulation technology are required to execute [7].

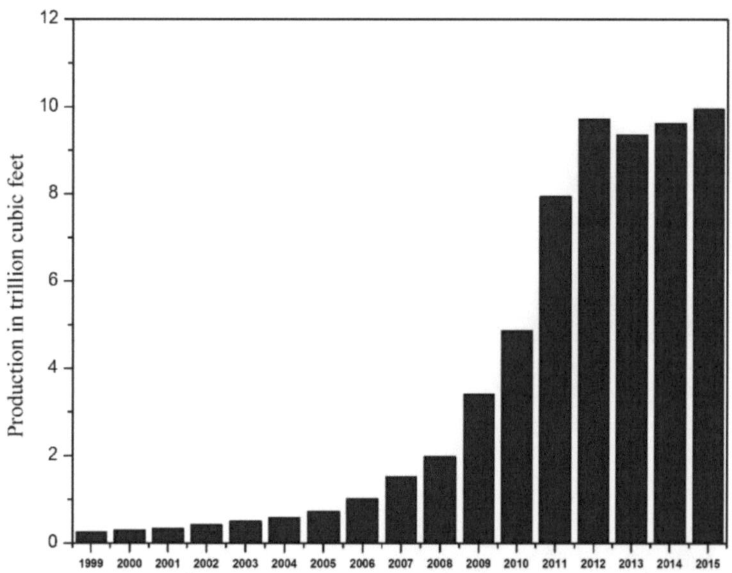

Figure 2.
Production of natural gas from shales in United State from 1999 to 2015 (in trillion cubic feet) [8].

2. Hydraulic fracturing technology

Hydraulic fracturing is classified as formation stimulation technique in which reservoir rock is fractured by pumping fracturing fluid with high pressure to create a fracture networks in order to increase the hydrocarbon production rate. This stimulation process has been used in 9 out of 10 gas wells in the United States [10, 11]. Water-based fracturing fluid has been widely used for fracking rock formation. In particular, shale reservoirs are characterized by their extremely tight rock formations with very low pore connectivity. The matrix permeability of a typical shale gas reservoir is about 1–100 nano-Darcy (nD) [12, 13], whereas the porosity is mostly less than 10%. To develop such ultra-low permeability formations, hydraulic fracturing is proven to be a successful method. In applications for shales, millions of gallons of water as the base fluid and sands in combination with a small amount of chemical additives are pumped into the reservoir [1, 14]. The injected fluid breaks the rock at high pressure and releases the free and adsorbed gas as shown in **Figure 3**. An immensely high permeability can be achieved by applying hydraulic fracturing that also aids in connecting the fracture networks [1]. During fracture generation and propagation, the sand or other coarse materials, the so-called proppant, is employed to expand the fractures as it holds the fractures open when the pressure is eventually relieved. The proppants can be classified into three types which are silica sand, resin coated sand, and ceramic proppant [15]. The utilization of proppant must be appropriate and its selection is strongly based on type and characteristic of well and reservoirs which will be hydraulically fractured. Proppant selection including its type, size, and shape is a critical element for the stimulation process whereby proppant characteristics such as weight, strength, consistency in size, and inert nature must be taken into account for effectively maintaining cracks from fracturing operation [16].

Furthermore, different types of fluids and treatments have been used and continuously developed for fracturing application. The effectiveness of fracturing fluids such as water, micellar solution, crosslinked-gel, polymer foam, and polymer-free foam has been studied and its selection is generally based on various factors including pressure gradient, reservoir temperature, formation Young's modulus, fracture half-length requirements, the presence of natural fracture, and

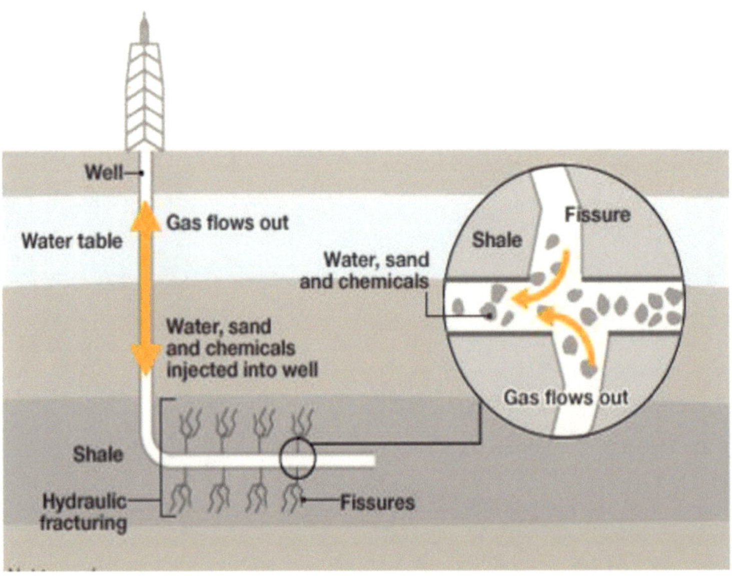

Figure 3.
Hydraulic fracturing of shale gas reservoir [1].

height of the pay zone. In addition to the aforementioned factors, chemical usage in the fracturing process is also adjusted depending on its environmental impacts and economics. On the other hand, there are several constraints associated with shale hydraulic fracturing processes that include usage of high volume of water, aquifer contamination, and methane infiltration in aquifers.

In general, fracturing fluid is composed of 99.5% of water and sand proppant, and 0.5% of chemical additives [17]. Potential additives for enhancing the performance of fracturing fluid include surfactant, friction reducers, biocides, and scale inhibitors [18]. In order to optimize fracture treatment for improving overall recovery efficiency, reservoir characterization is essential as specific treatment is required depending on the rock and fluid properties. Besides hydraulic fracturing, other techniques such as explosive fracturing and dynamic loading that do not utilize water-based fluid have been taken into consideration [19]. Nonetheless, these aforementioned techniques are not extensively implemented due to their performance and environmental concerns.

3. Fracturing fluids

The formation of fractures in reservoir rocks is initiated by fracturing fluid injection under high pressure to hydraulically break the rock, hence producing the stored hydrocarbons. Fracture treatment and fracturing fluid design are essentially dependent on the unique properties of reservoirs. Alteration of fracturing fluids is important in order to meet the targeted reservoir and operating conditions. Oil-based fracturing fluids were initially developed for fracturing job; however, due to the environmental and safety concerns, it was shifted to water-based fracturing fluids. An excessive amount of water utilization which can cause damage to water-sensitive formations has led to the use of liquefied natural gas as an alternative. Besides the use of slickwater, chemical solutions for hydraulic fracturing have also been known as an effective technique for complex reservoirs which are naturally fractured, brittle, and tolerant of high water volume [20]. Other innovations such as water-based viscous polymeric fracturing fluids have been proposed, but they are still associated with some challenges, such as degradation of different molecular weight polymers and the formation of internal filter cake leading to undesirable damage to the reservoir rocks [20]. Slickwater which is mainly composed of water with a low concentration of chemical additives, or combination of different fracturing fluids has been commonly used for shale gas wells. As mentioned earlier, owing to different purposes of fracturing jobs, the utilization of other additives including acid, surfactant, potassium chloride, friction reduces, corrosion inhibitors, and pH adjusting agent at low concentration has been considered [21, 22].

3.1 Hydraulic fracturing fluids for shales

Shales have great variations with their typical characteristics that essentially determine the required hydraulic fracturing technique and fracturing fluid design. For shale fracturing jobs, fracturing fluid comprises of base fluid, additives, and proppant. Slickwater treatments using high injection rates and lower proppant concentrations have provided some advantages such as lower cost, reduced fracture height growth, and reduced gel damage within the fracture. However, the use of high volumes of fluids, poor proppant transport and suspendability, higher leak-off, and low fluid viscosity causing complex fracture geometries are disadvantages associated with slickwater usage as fracturing fluids [21]. Surfactant-based fluids were then proposed as fracturing fluid because surfactant molecules can undergo self-association to generate micelles that can increase viscosity in the absence of polymer. Some modification of surfactant-based fluid system, such as nanoparticle addition, has been considered to stabilize the system at high temperature [23].

Furthermore, single gas system has been reported to effectively increase the amount of energy required to recover the fluid and reduces water volume in shales that are classified as water-sensitive zones. However, there is a major disadvantage associated with the gas fracturing process which is the reduction in the amount of proppant that can be placed. The cost of operation also increases due to capturing, pressurizing, and transportation of the gas, for instance CO_2. Additionally, separation of CO_2 and CH_4 during flow back would need additional facilities which will increase the expenses and the produced natural gas along with CO_2 during the flow back period will also reduce. Supercritical CO_2 has the ability to dissolve some amount of the water formed. When the amount of water reduces in order to achieve equilibrium with supercritical CO_2, the remaining super saturated brine would cause salt precipitation which could block the flow channels and restrict the production [24].

In water-sensitive reservoirs possessing high clay contents, fracturing fluid containing a small amount of water and large gas volume is preferred in order to reduce formation damage caused by high capillary pressure and permeability discontinuity as the impacts of clay swelling [25]. Foam also can reduce the damage around wellbore due to invaded fluid which eventually reduces the water volume used for hydraulic fracturing. Mixture of dispersed gas (N_2 or CO_2) and surfactant solution resulting in foam system has become another innovation whereby the foam can carry proppant efficiently with minimum residue left in the fracture. This system, i.e., N_2 foam or CO_2 foam, which is also known as energized fluid, has higher propagation ability into more complex fracture networks due to its mobility control ability. In other words, foam is able to carry proppant deeper in the formation in more efficient manner. CO_2-based energized fluids have been reported to provide a better foaming performance leading to a higher recovery [26]. In ductile reservoirs, an efficient proppant placement is essentially required and fracturing techniques such as N_2 foam and CO_2 polymer have been implemented in ductile reservoir, e.g. Montney Shale in Canada [27]. CO_2-based fluids can eliminate the need of water, provide extra energy due to gas expansion, and help in decreasing the flowback time.

Recent developments of unconventional reservoirs including shale and tight gas and coalbed methane have put more emphasis on fracturing treatment with little use of water as the interaction between these reservoirs and the used fracturing fluids can negatively impact gas production [28]. The attempt to reduce water use in the fracturing process has been driven by several factors explained below in detail.

3.1.1 Water-sensitive formations

The recovery of water, oil, and gas from unconventional reservoirs is essentially affected by the mineralogy of the rock formation. Ultra-tight formation with small propagated and natural fracture widths results in high capillary forces which are important for hydrocarbon production. The injection of water causes capillary barrier leading to production decline. In the case of water-sensitive formations having high clay content, clay swelling occurs during fracturing processes with water-based fracturing fluid which can reduce formation permeability due to peeled pore surface and pore throat plugging. Changes in permeability due to clay swelling lead to capillary pressure and relative permeability shifts. These effects become more dominant when moving from micro to nano-Darcy permeability ranges [21, 29]. The excessive fine migration including clays in the near-wellbore region can also reduce the productivity. To avoid clay swelling and fine migration, different fracturing treatments utilizing a little amount of water, such as oil-based fluids, high quality foams, and liquefied petroleum gas are preferred.

3.1.2 Water blocking

During fracturing fluid injection, water invasion occurs in the formation due to high capillary forces. High surface tension in pores and the hydrophilic characteristic of rock surfactant result in a significant water block in the formation, illustrated in **Figure 4**. This effect will be detrimental to gas productivity [30]. The increase of water saturation due to water trapping or water blocking considerably decreases the relative permeability of the gas [31].

As compared to conventional fluid, CO_2-based fluid can assist the clean-up of injected liquid phase during flowback. As the pressure decreases, the expansion of the gaseous phase assists the flow to the surface and provides a rapid fracture and reservoir clean-up which accelerates the onset of the production phase after speeding up the flowback phase [30].

3.1.3 Proppant placement

The slickwater/water-based fluid fracturing process generally creates longer and skinny fractures. However, a poor delivery of proppant has been reported along much of the fracture and much of the fractures remain un-propped allowing it to close after the pressure is released and fracturing job is over, particularly in ductile rock formation [32]. In addition, the near wellbore region, in this process, is dominantly propped due to rapid sand settling. Gels are used to avoid this rapid settling of proppants; however, the adverse effect which is damaging the proppant pack and the fracture surface can occur. In previous studies, it was suggested to replace sands with ultra-lightweight proppants in order to achieve efficient transport of proppant. Foam fracturing treatment reportedly gives an efficient transport of proppant as compared to the slickwater fluid treatment. Therefore, use of foams can be considered to effectively improve the performance of proppant placement. In foam fracturing, water exposure is avoided alienating the reservoir matrix from the softening effect and hence proppant embedment could be reduced [30].

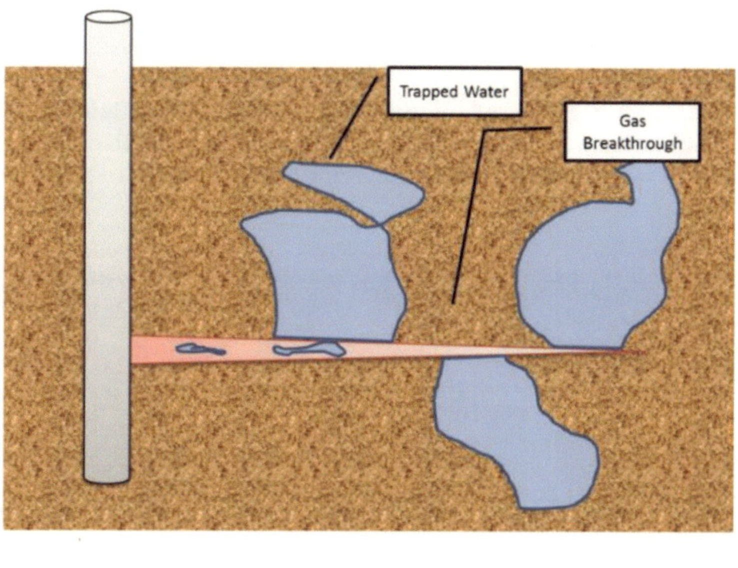

Figure 4.
Water blocking due to high capillary pressure.

Moreover, the interaction between bubbles of gas at high foam qualities gives a large energy dissipation which results in good effective viscosity hence providing more effective proppant placement. Whereas at low foam qualities, the interaction among gas bubble is minimum and the fluid viscosity behavior is similar to that of base fluid. **Figure 5** shows the condition of proppant pack formed using three different fracturing fluid systems [33]. It is clear that energized fluid or CO_2 foam provides efficient proppant placement, whereas the water and gel-based proppant have poor permeability due to proppant embedment and gel residue, respectively [33].

3.1.4 Water availability and cost

The amount of water required for the fracturing process depends on the type of formation which is being fractured. The utilization of a huge amount of water is a major concern and the equipment and the fluids are limited on the fracturing site. The widely used fracturing technique employs water-based fluid which consumes a large amount of water, has high water leakage in formation, and imposes high water disposal costs. Besides, the cost and supply of fluids such as LPG, N_2, and CO_2 highly depend on the field location.

3.2 Foam fracturing fluid

Foams have been found to be the most promising and appropriate fluid to fracture shales and improve the recovery efficiency. Although the cost of operation increases, the benefits are much higher than the incremental cost [35]. The structure of foams has the ability to provide an increased effective viscosity without plugging reservoir pores and causing formation damage by forming any filter cake [28]. It has an increased efficiency due to reduced fluid loss coefficients, high viscosity inside induced fractures and negligible sand settling velocities [28, 36]. Foam application also gives an increased capability of proppant distribution and proppant placement over the entire fracture length. Due to high foam apparent viscosity, it is achievable to have an improved proppant suspension and placement. In foam fracturing, the utilization of gas as a replacement to a significant amount of the liquid phase assists hydrocarbon recovery by decreasing formation damage and water blocking. Foam utilization eliminates the need of any additional additives such as cross linkers, gel breakers, etc. It also decreases the amount of produced water and its treatment cost. Moreover, the expansion of gas assists liquid flow back and helps fracture cleanup.

Foams are typically generated by a surfactant solution (base fluid), in some cases, in combination with a small amount of polymer as a stabilizer and other additives. Surfactants that are used as a foaming agent may help to lower the surface

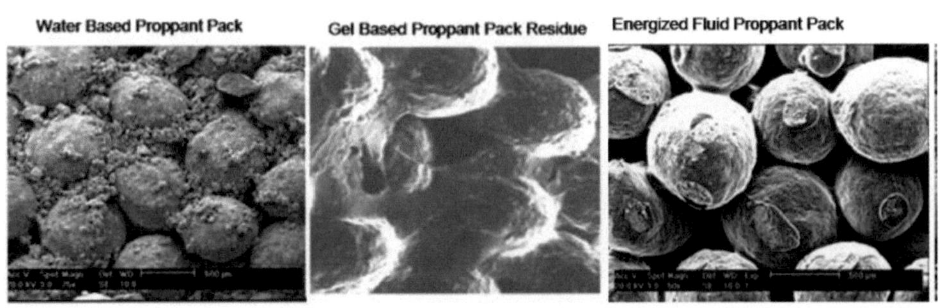

Figure 5.
Proppant conditions after performing fracturing job using different types of fracturing fluids [34].

tension of the fracturing fluid and avoids water blocking by recovering fracturing fluid after the job completion [37]. Both laboratory and field scale tests have shown that the addition of surfactant increases the gas production by reducing the capillary forces and altering the wettability of shales. The rheology of foam has great importance in fracture treatment design and has been discussed by multitude of studies [8, 36, 38–40].

4. Foam rheology

Determining foam rheology is complex and it is considered difficult to predict the behavior of foam flow [20]. The performance of foam-based fracturing job is highly dependent on the rheology of the foam under downhole conditions and the efficiency of fracturing job depends on non-Newtonian behavior of foam [41, 42]. Foams are considered versatile, complex, and unique due to their high viscosity and low density characteristics [43]. Foam apparent viscosity is determined by accounting for the contribution of foam film thickness, bubble deformation, and the expansion of foam interface due to surface tension gradient [44]. The apparent viscosity of foam is strongly dependent on various process variables such as foam quality, shear rate, temperature, pressure, surfactant concentration, and salinity [8, 41, 42]. The effects of these parameters are discussed below.

4.1 Effect of foam quality and texture

Foam quality and its texture have a strong impact on the viscosity of CO_2 foam [45, 46]. A simultaneous study of foam texture is important during the foam rheology measurements [47–49]. Before starting the viscosity measurements, it is important to ensure that the flow loop is completely filled with the foam of known foam quality. It has been reported by many researchers that the rheology of foam is dependent on foam quality and foam apparent viscosity [38, 48, 50–53]. Foam quality (f_q) is defined as the volume fraction of gas in foam [54–56] and is expressed as Eq. (1).

$$f_q = \frac{V_g}{V_g + V_l} \tag{1}$$

where v_g and v_l are gas and liquid volume in foam, respectively.

When the foam is relatively wet, i.e., at low quality, the foam bubbles are less in number and are far apart with no interaction with each other during the foam flow. Therefore, the foam viscosity is low.

When the foams have low foam quality (i.e. relatively wet foam), the interaction of dispersed gas bubble is insignificant during the foam flow and due to this reason viscosity of foam decreases ; whereas at high foam quality, i.e., when the foam is dry, the interaction of bubble will be quite significant during the foam flow and the friction between the individual bubbles will result in the drastic increase in foam apparent viscosity. In the case of dry foam with very high foam quality, the bubble cannot sustain and breakdown occurs during which a sharp decrease in foam viscosity occurs [43].

4.2 Effect of shear rate

The applied shear rate has a significant impact on foam apparent viscosity. The changes in the foam apparent viscosity at different shear rate display power law behavior. Ahmed et al. [84] tested high quality polymer free foams and reported a typical shear thinning behavior within the shear rate range (10–500 s^{-1}). The viscosity of foam decreases in shear flow due to Rayleigh-Taylor instability, which causes the tensile deformation, stretching, and rupturing of lamella [41, 57].

4.3 Effect of temperature

Foam is a metastable state system which goes through coarsening and rupturing when the liquid drains from the lamella and plateau borders [58]. An improved stability foam could be achieved when the viscosity of foaming solution is increased [59]. The increase in temperature causes thermal thinning of foam film which quickly drains liquid leaving behind thin lamella [60]. **Figure 6** shows the foams generated using a mixing approach at two different temperatures (20 and 50°C). It is clear that the foam texture at high temperature is relatively coarser and it has a wide range of bubble size distribution whereas at low temperature, uniform and fine textured foam is noticed. Hence, when the temperature is increased, the rate of foam lamella drainage and coalescence of bubbles are quick, resulting in a significant decrease in foam apparent viscosity [8, 60–62]. High fluctuations in temperature may form holes in the lamella which would increase both bubble coalescence and lamella rupturing [62]. Hence, the rising temperature causes significant reduction in the apparent viscosity of supercritical CO_2 foam [8, 63].

4.4 Effect of pressure

When the pressure increases, the size of foam bubble significantly decreases, whereas the lamella size becomes thinner and larger which results in slow liquid drainage [59]. This is due to the reason that the generated foams at high pressure are exceptionally strong possessing high apparent viscosity [8] and if the pressure is extremely high, it may be possible that the lamella could not withstand and it ruptures [59].

4.5 Effect of surfactant concentration

Proper selection of foaming surfactant and its concentration under reservoir and fracturing conditions is an essential task. Aronson et al. discussed the disjoining isotherm of foam film at two different surfactant concentrations

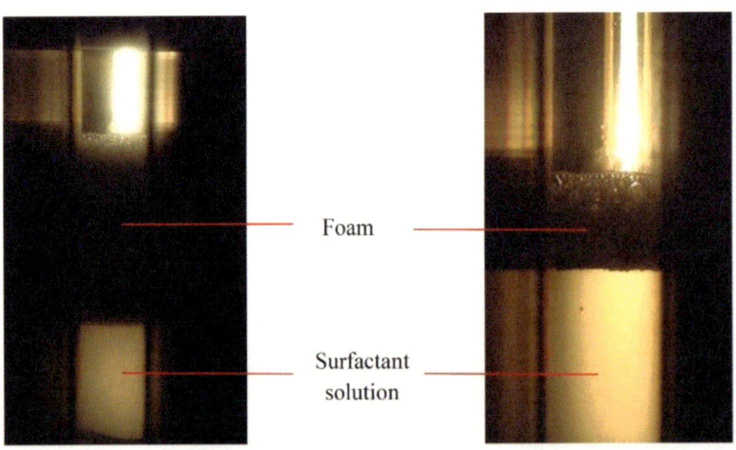

Figure 6.
Pictures of foam column stabilized by surfactant (Triton X-100) at 20°C (left) and 50°C (right) [64].

[65]. They found that higher surfactant concentration is able to provide high disjoining pressure to the foam lamella which significantly increases the pressure gradient and the resistance to the foam film during the foam flow process. Apaydin and Kovscek discussed the impact of surfactant concentration and noticed that at low surfactant concentration, weak foam generates which offers low resistance to flow [66]. Gu and Mohantay discussed the apparent viscosity of polymer free foam considering two different surfactant concentrations (0.1 and 0.5 wt%) and it was noticed that the higher concentration is able to generate highly viscous foam under fracturing conditions [8]. The increase in viscosity is accredited to the increment in total interfacial area of the foam structure which induces additional lamella stability [8]. Foam lamella should be elastic in order to withstand any deformation and the force which restores lamella comes from the Gibbs-Marangoni effect [67]. Some authors presented that direct and strong relationship exists between surface elasticity of lamella and foam stability [68–72], whereas others disagree and reported no direct relationship [67, 73]. A maximum foaming performance is usually noticed at intermediate surfactant concentration dictated by the Gibbs-Marangoni effect [69]. When the surfactant concentration is high, the surface elasticity of lamella decreases, which negatively affects the foam endurance due to reduced counteraction towards the deformation forces [69].

4.6 Effect of salinity

The shape of surfactant micelles (known as aggregates of surfactant molecules) changes with the change in surfactant packing parameter. The packing parameter is expressed as $P = v/a_0 l_c$, where a_o is the area of the surfactant headgroup, v is the volume of the surfactant tail, and l_c is the tail length of the surfactant molecule [74]. Due to the increase in salinity, the transformation of spherical shape micelles of ionic surfactants into wormlike micelles takes place. These micelles are elongated spherocylindrical having two hemispherical end caps and a cylindrical body. The neutralization of repulsive forces between micelles due to the addition of salt reduces the effective area of surfactant head and alters the packing parameter of surfactant [14, 75]. The wormlike micelles entangle with each other and generate three dimensional networks, which impart viscoelasticity, and therefore, the behavior of surfactants becomes similar to that of viscoelastic polymer solutions [74]. Anionic surfactants have the ability to form such wormlike micelles when sufficient electrolyte is added [76]. For surfactants, such as sodium lauryl ether sulfate (SLES) and sodium dodecyl sulfate (SDS), the addition of electrolyte increases the surfactant packing parameter as well as solution viscosity [74]. It has also been reported that surfactant ability to form a strong foam depends on its hydrophilic/lipophilic balance (HLB) which may vary due to the addition of salinity [77]. Foam of decyltrimethylammonium bromide (DTAB) surfactant has been reported to decrease the foam viscosity with the increase in salinity; however, the cetyl trimethylammonium bromide (CTAB) provided an increasing trend as the salinity in the solution was increased. Another surfactant, Mackam CB-35, by Rhodia provided a decrease in foam viscosity until 3 wt% salinity, and beyond that a prominent increase in foam viscosity was reported [77].

4.7 Foam rheological models

The rheology of foam determines various characteristics of fracture growth, and therefore, it is important to accurately estimate the rheology in order to predict the fracture geometry. Different rheological models have been developed that describe the foam flow behavior, from the widely used power law model to the Herschel-Bulkley model, which have different degrees of success [8, 32].

The Herschel-Bulkley model incorporates a yield stress for non-Newtonian fluid. In this model, the yield stress becomes negligible at high shear rate and the model becomes similar to the power law model. Mathematically, the Herschel-Bulkley model is presented as Eq. (2) [8].

$$\tau = \tau_o + K\gamma^n \tag{2}$$

where τ is the shear stress, γ is the shear rate, τ_o is the yield stress, K is the consistency index, and n is flow behavior index.

Power law or Ostwald-de Waele model is one of the most commonly used models for describing the non-Newtonian behavior of foam [8, 63, 78–80]. The power law model can be mathematically expressed as shown in Eq. (3) below [63, 81].

$$\mu = K\gamma^{n-1} \tag{3}$$

where μ is the viscosity, K is the flow consistency index, γ is the shear rate, and n is the flow behavior index.

A straight line appears when log μ is plotted versus log γ. By taking the logarithm of both sides of Eq. (5), the parameters of the power law model can be determined as shown below.

$$\log\mu = \log K + (n - 1)\log\gamma \tag{4}$$

If the solution viscosity of the solution is plotted against the corresponding shear rate on a log-log paper, a straight line appears with the intercept as K at a shear rate (1/s), and as shown in **Figure 7**, $(n - 1)$ will be the slope of the straight line.

The n value explains the behavior of the solution, i.e., when $n < 1$, the fluid shows shear thinning behavior, whereas for the shear thickening fluids, $n > 1$. For foams, $n < 1.0$ indicates a pseudoplastic behavior. The extent of shear thinning behavior of solutions can be quantified by the value of n. The value of n is significantly lower than unity if the solution is highly shear thinning, whereas if the n value is equal to unity which is the case of Newtonian fluid, the K value will become the Newtonian viscosity.

Foam behavior indices (K and n) are the function of foam quality, chemical concentration, temperature, and pressure. The rheological behavior of the foam is somewhat similar to the polymers. Foam system is considered complex and its model parameters are reliant on foam geometry, temperature, pressure, and foam properties [8, 41, 78]. Previous studies performed on foam rheology concluded that it is important to control various parameters such as gas volume fraction (i.e. foam quality), foam texture, pressure, temperature, chemical types, concentrations, etc. while measuring the foam apparent viscosity [82]. Many studies also reported higher performance of CO_2 foam fluid with higher recoveries as compared to other fluids [21, 83]. However, it is difficult to understand and model the behavior of such energized fluid [83].

Ahmed et al. investigated the effect of various process variables such as pressure, temperature, salinity, surfactant concentration, and shear rate on CO_2 foam apparent viscosity under high pressure high temperature conditions and presented a set of empirical correlations [39, 84]. In their study, the polymer free foam was generated using a conventional surfactant, i.e., alpha olefin sulfonate (AOS) and a

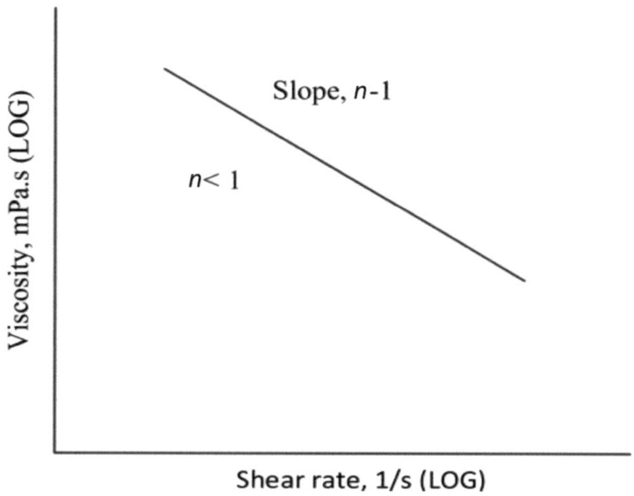

Figure 7.
Power-law model [8, 78].

foam stabilizer, and it was noticed that all the aforementioned process variables are strongly dependent on foam apparent viscosity (discussed above in Section 4). All the foams exhibited a typical shear thinning behavior within the tested shear rate range (10–500 s^{-1}) and the power law model was fitted on experimental data. They presented a set of empirical correlation to predict apparent viscosity of CO_2 foam. Power law indices were found to be strongly dependent on all process variables. The new equations for K and n were developed as a function of process variables which were then substituted in the power law model. These correlations could cover a wide range of conditions and were found accurate in predicting the viscosity of CO_2 foam fracturing fluid. These developed models could be integrated into any fracturing simulator in order to evaluate the efficiency of foam fracturing fluid.

Gu studied foam fracturing using polymer free foam considering ultra-lightweight proppants (ULWPs) [8, 32]. They also developed empirical correlations through the modification of the power law model, which were then applied in a fracturing and reservoir model using a commercial simulator CMG IMEX. They used ULWPs to predict the formation productivity with both slickline and polymer free foams. They have been able to present foam-based hydraulic fracturing fluid which has efficiently propped the fractures and utilized less water compared to that of slickline fluid. Furthermore, he evaluated the designed foam fluid and proppant using a combined experimental and computational modeling technique which helped in identifying the optimal proppant amount and gas liquid fraction (or quality) of foam.

4.8 Experimental study of foam rheology

Numerous experimental studies used pipes with a small diameter to investigate foam rheology. This is a more reliable method of studying foam behavior in wellbores. Foam deteriorates due to its unstable nature which is caused by liquid drainage under the action of gravity [85]. Accumulation of the liquid takes place at the bottom of the samples and foam cannot be taken as a homogeneous system. Foam is made to flow through a steel recirculation loop in which pressure drop over a certain length is measured and apparent viscosity was calculated. Hagen-Poiseulle equation is used to compute the apparent viscosity of foam in pipe or tubing and it is represented in Eq. (5) [41, 53, 57, 74, 77].

$$\mu_{app} = \frac{D^2 \Delta P}{32LU} \tag{5}$$

where μ_{app} is the foam apparent viscosity, D is the diameter of tubing, ΔP is the differential pressure between the test sections, L is the tubing length, and U is average velocity determined from the total volumetric flow rate of foam.

Before carrying out any measurements, a constant shear rate needs to be set to ensure uniformity across the foam. Once the foam is equilibrated in the recirculation loop, the pressure drop is measured rapidly at different flow rates while ensuring that the foam texture does not vary over time. Patton et al. measured foam viscosity as a function of shear rate using a viscometer apparatus [86]. They mixed constituents of the foam and passed it through the foam generator, i.e., packed bed. Flow rates, temperature, and pressure drop were measured after displacing the foam through the small diameter tube [87].

Xue et al. [74], Li et al. [57], and Sun et al. [41] recently used a flow loop system for fracturing foam studies by using CO_2 foam. These rheology studies involved foam as fracturing fluid and the investigation was made at downhole condition using a flow loop system. The effects of temperature, pressure, foam quality, and shear rate on fracturing foam were studied. Sudhakar and Shah (2002, 2003), Bonilla and Shah [88], and Sani et al. (2001) used a recirculation loop rheometer to investigate the rheology of polymer foam and the power law behavior was observed [88]. These experiments determined that foam behaves as a non-Newtonian fluid. The foam apparent viscosity decreases as the shear rate increases and such behavior is termed as pseudoplastic.

5. Foam fracturing fluid performance: laboratory studies

5.1 Foaming agent and water recovery

Prior to the implementation, it is important to investigate the performance of fracturing fluid at reservoir conditions. The measurement of fundamental properties of the used foaming agent such as interfacial tension and contact angle that are the basis for reducing capillary pressure are also essential to perform [89, 90]. Additionally, the adsorption of surfactant as a foaming agent is equally important to study in order to estimate the chemical loss in the reservoir during the recovery process. The recovery of injected fluid as well as chemical performance can be experimentally evaluated based on core flow tests. Furthermore, it has been reported that several field pilot tests were conducted to check the performance of surfactant for fracturing applications. Most of the pilot tests were conducted in the Barnett and Marcellus shales utilizing conventional surfactants which include nonionic alcohol ethoxylate surfactants and amphoteric and cationic surfactants [89]. The water recovery using conventional surfactants has been reported to achieve approximately 60% (an average of 3 wells) [89]. Barnett shale is considered notorious for retaining water. Another case study was conducted considering a conventional surfactant and only 2300 bbl of water was recovered out of injected 6430 bbl giving about 28% recovery [89]. A nanofluid has also been employed in fracturing job to reduce the chemical adsorption whereby the recovery of injected water reached about 40% in this case [91, 92]. Both laboratory and field studies revealed that the addition of surfactants to the fracturing fluid system helps in increasing the recovery of additional water. Besides, an increase in overall gas production was also observed.

5.2 Foam screening and optimization

The proper selection of foaming agent for generating foam fracturing fluid is required to optimize the process under desired conditions. Lin et al. investigated different surfactants at high temperature and the results were compared with those of a conventional foaming agent used previously as fracturing fluid [93]. They performed surface tension, foam stability, and foam viscosity experiment in order to evaluate the surfactant performance for foam generation. It was ascertained that the results from foam stability experiment could give the indication of surfactant performance for foam generation but the foam behavior under fracturing conditions could not necessarily be deduced. The utilization of flow loop foam rheometer has enabled the study on foam viscosity under reservoir operating temperature and pressure conditions. Based on extensive laboratory studies which mainly include foam rheology, a superior performance foam formulation was obtained which was able to provide a stable foam with both CO_2 and N_2 in a high temperature environment especially where other conventional foamers failed to generate the stable foam. The selected foamer was also found to be having good compatibility with other chemicals in fracturing fluid systems and it has also provided low emulsification as compared to other foamers. When the selected foam formulation through this detailed screening and optimization procedure under reservoir conditions was employed in various fracturing treatments in fields, successful results were achieved.

5.3 The role of chemical additives

Many attempts have been carried out to find the formulation of foam fracturing fluid that can meet the requirement of targeted reservoirs and provide an optimum performance. The synthetic polymer with high concentration has been reported to meet the need of higher viscosity of fracturing fluid. However, the increase in polymer loading would give more severe formation damage due to the formation of fluid residue.

Some viscoelastic surfactants (VES) have been proposed to generate highly stable and viscous foam fracturing fluid with minimum water contents at high temperature conditions. The increase in surfactant concentration may form worm-like micelles which impart viscoelastic property to the foam lamella, hence generating high strength foam which is a relatively clear approach as compared to that of polymer solutions. The addition of nanoparticles (zinc oxide (ZnO) and magnesium oxide (MgO)) has also been found to improve the temperature tolerance of this type of surfactant from 93 to 121°C [94].

The evaluation of different types of polymer as additives for achieving stable and viscous foam has also been performed by Ahmed et al. [38]. They utilized both conventional HPAM polymers and new associative polymers (possessing higher level of hydrophobes) for bulk foam stability tests using FoamScan as well as foam viscosity measurement using HPHT foam rheometer. It was concluded that the use of conventional polymers was not preferable under harsh reservoir conditions as they are more prone to degradation instead of stabilizing the foam. Meanwhile, associative polymer provided dramatic increase in the viscosity and stability of CO_2 foam without undergoing degradation under testing conditions.

More complex system of CO_2 foam has been presented by Xue et al. in which the foam was stabilized with betaine surfactant and silica nanoparticles in the absence and presence of synthetic polymer [95]. The texture of foam was observed from the view cell connected to foam generator having glass bead pack with a permeability of 23 Darcy and the viscosity of foam was measured using a capillary viscometer. It has been reported that an extremely dry (90–98% foam quality) supercritical CO_2-in-water-foam prepared in brine presented a high viscosity and stability at 50°C. The data on such high quality foams are very limited [95, 96] and the proposed combination (i.e. surfactant/nanoparticle/polymer) has indicated the possibility of having the strong foam with minimum water content which is desirable for fracturing application. Lauryl amidopropyl betaine, the used surfactant, was able to lower the CO_2/water interfacial tension to 5 mN/m. This surfactant attracted anionic NPs (and anionic HPAM in the systems containing polymer) to the CO_2/water interface. Nanoparticle presence in the CO_2 foam system has remarkably slowed down the Ostwald ripening phenomena, whereas, the polymer slowed down the liquid drainage by increasing the viscosity of continuous phase and surface viscosity [95].

5.4 Consideration of different process variables

The rheological properties of foam as fracturing fluid have been evaluated by Wilk et al. using flow loop rheometer with a purpose of reducing the amount of water during fracturing operation [97]. They reported that the concentration of foaming agent and polymer prominently impact the foam rheological parameters. The evaluation in rheometer was performed on the basis of two different foam qualities, i.e., 70 and 50%. Their results showed that the 70% quality foam was able to provide stronger foam due to favorable changes in the foam structure and distribution. The foam texture study also revealed that the bubble sized distribution was altered due the presence of different additives and in this case, polymer additive was able to further reduce the bubble diameter which resulted in high viscosity foam.

Ahmed et al. has investigated polymer free foam performance with different process variables such as surfactant concentration, salinity, temperature, pressure, and shear rate utilizing a high pressure high temperature flow loop system [39, 84]. The rheology of CO_2 foam was presented under a wide range of different parameters such as temperature (40–120°C), pressure (1000–2500 psi), salinity (0.5–8 wt%), surfactant concentration (0.25–1 wt%), and shear rate (10–500 s^{-1}). They presented that 80% foam quality which was found to provide the best performance, expected to be favorable for fracturing job at targeted conditions. These experimental studies found that the foam apparent viscosity under supercritical conditions considerably increases with the increase in surfactant concentration up to 5000 ppm, whereas a continuous increasing trend in foam viscosity appeared as the salinity in the foaming solution was increased. Besides that, the increase in temperature resulted in thermal thinning of foam lamella whereby dropping the foam viscosity. Also, the increment in pressure provided stability to the foam lamella which resulted in high viscosity foam. It was reported that all the foams behaved as shear thinning fluid within the tested shear rate range and power law was fitted on foam viscosity data. The flow behavior indices were found to be the strong function of aforementioned process variables and using viscometric data, the new empirical correlations have been developed, which gives accurate prediction of the apparent viscosity of CO_2 foam as a function of process variables. These empirical models may help in predicting the optimum fracturing conditions and also for the fracture simulation modeling study.

5.5 Proppant transport effectiveness

The experimental study in Hele Shaw slot (**Figure 8**) was presented by Tong et al. for visualizing the proppant transport behavior during foam injection [98]. The foam was generated using conventional surfactant in combination with low molecular weight (8 million Da) polymer solution as viscosifier. In their study, static

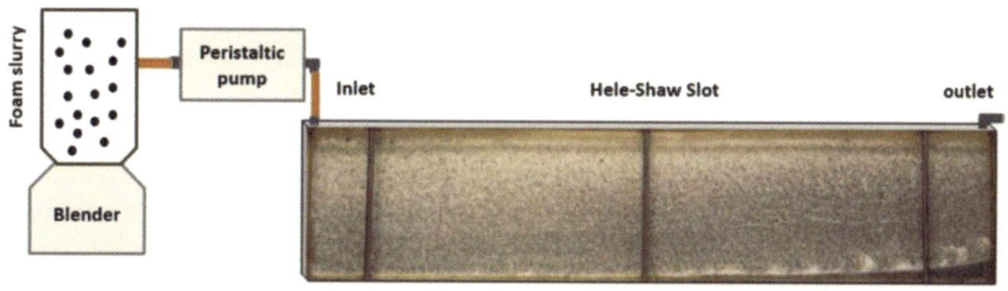

Figure 8.
Laboratory scale Hele-Shaw fracture slot for mimicking induced fracture [98].

foam stability and foam viscosity tests were conducted to study the durability and strength of the generated foam. The visualization of foam-based fracturing fluid was studied at different foam qualities and it was reported that proppant transport was quite efficient with high quality foam due to its high viscosity and stability. They reported that the low quality foam was not an effective proppant carrying medium due to high liquid drainage and low apparent viscosity of foam during its transport through the fracture system. It has also been reported that the dry foam allows extremely slow proppant settling compared to wet foams even during high proppant loading.

6. Conclusions

Advanced horizontal drilling and hydraulic fracturing have been widely applied for unlocking a huge amount of hydrocarbon reserves from tight shales. Conventional water-based polymeric solutions have been commonly utilized for fracturing jobs due to their ability to transport proppants deep into the reservoir. However, the use of polymer tends to cause plugging of nanopores and detrimentally affects the shale productivity. Moreover, many environmental constraints are associated with the use of a huge amount of fresh water and the disposal of contaminated water during flow back period is continuously reported, driving a need of waterless fracturing fluid system.

CO_2 has been known as alternative fracturing fluid with various added benefits including releasing the adsorbed gas, water flow back improvement, and carbon sequestration. However, the use of single CO_2 system has limitations affected by the low viscosity of gas, limited proppant carrying ability, and limited possibility to operate at depth. An innovation to have waterless fracturing fluid has also been attractively developed resulting in less water consumption, less formation damage, and less liquid to recover during the fracturing process.

The combination of surfactant with CO_2 generates foam which is considered as a highly attractive and unique solution to all the above associated concerns during fracturing operation. CO_2 foam has high viscosity, good thermal stability, better proppant transport and placement ability, stable rheological performance as compared to polymers, ability to reduce clay swelling and fine mitigation issues, and increased flow back due to gas expansion. Additionally, surfactants in the base fluid reduce capillary forces and alter shale wettability, which assists water flowback, and increases gas production. Therefore, selection of an appropriate surfactant is of prime importance. However, available literature studies on the evaluation of surfactant and foam performance as fracturing fluid are limited.

In spite of all exceptional benefits, a good understanding of foam rheology is required for the design of optimum foam fracturing treatment. The foam fracturing process highly depends on foam viscosity and it is highly desirable that the foam should provide sufficient viscosity for efficient job completion under reservoir design and operating conditions. According to the previous studies explained in this chapter, besides the evaluation based on foam stability, an analysis of the applicability of foam-based fracturing fluid could be derived from several experimental investigations including foam viscosity measurement using flow loop foam rheometer which also could provide the information of foam texture at different foam qualities and fracturing conditions. A thorough screening and optimization of foam considering different variables under fracturing conditions could effectively improve the efficiency of fracturing job. Studies also have implied that the foam rheological property is challenging to estimate due to numerous variables involved.

Conflict of interest

The author declares no conflict of interest.

Author details

Shehzad Ahmed[1*], Alvinda Sri Hanamertani[2] and Muhammad Rehan Hashmet[3]

1 Ali I. Al-Naimi Petroleum Engineering Research Center (ANPERC), King Abdullah University of Science and Technology, Thuwal, Saudi Arabia

2 Department of Petroleum Engineering Universiti Teknologi PETRONAS, Seri Iskandar, Malaysia

3 School of Mining and Geoscience, Nazarbayev University, Astana, Kazakhstan

*Address all correspondence to: shehzadahmed@kaust.edu.sa

References

[1] Lee KS, Kim TH. Integrative Understanding of Shale Gas Reservoirs. Switzerland: Springer; 2016

[2] Aguilera R. Flow units: From conventional to tight-gas to shale-gas to tight-oil to shale-oil reservoirs. SPE Reservoir Evaluation & Engineering. 2014;**17**:190-208

[3] Cander H. PS what are unconventional resources? A simple definition using viscosity and permeability. In: A.A.o.P. Geologist, editor. AAPG Annual Convention and Exhibition. Tulsa, US: American Association of Petroleum Geologists and Society for Sedimentary Geology; 2012

[4] Hinkley R, Gu Z, Wong T, Camilleri D. Multi-porosity simulation of unconventional reservoirs. In: SPE Unconventional Resources Conference Canada. 2013

[5] Harvey T, Gray J. The unconventional hydrocarbon resources of Britain's onshore basins–shale gas. Rep. Dept. Energy Climate Change. Report, United Kingdom. 2013

[6] Goodwin MJ, Musa OM, Steed JW. Problems associated with sour gas in the oilfield industry and their solutions. Energy & Fuels. 2015;**29**:4667-4682

[7] EIA. Technically Recoverable Shale Oil and Shale Gas Resources: An Assessment of 137 Shale Formations in 41 Countries Outside the United States. DC, US: U.S. Department of Energy, Washington; 2013

[8] Gu M, Mohanty K. Rheology of polymer-free foam fracturing fluids. Journal of Petroleum Science and Engineering. 2015;**134**:87-96

[9] Sun Y, Ok JT, Wu Q, Xiao F, Elgmati M. Using Single-Molecule Imaging System Combined with Nano-Fluidic Chip to Understand Fluid Flow in Tight and Shale Gas Formation Project Number: 09122-29. 2012

[10] Yongnian Z, Tong SY. China's Evolving Industrial Policies and Economic Restructuring. New York: Routledge; 2014

[11] Craig RK. Hydraulic fracturing (fracking), federalism, and the water-energy Nexus. Idaho Law Review. 2012;**49**:241

[12] Williams KE. PS the Permeability of Overpressure Shale Seals and of Source Rock Reservoirs is the Same. California, United States: American Association of Petroleum Geologists; 2012

[13] Lee SJ, Kim TH, Lee KS. Pressure transient characteristics of a fractured horizontal well in shale gas reservoirs. In: The Twenty-Third International Offshore and Polar Engineering Conference. 2013

[14] Feng Y, Chu Z, Dreiss C. Smart wormlike micelles. Chemical Society Reviews. 2013;**42**:7174-7203

[15] Yew CH. Mecânica do Fraturamento Hidráulico. Brasil: Editora E-Papers; 2008

[16] Liang F, Sayed M, Al-Muntasheri GA, Chang FF, Li L. A comprehensive review on proppant technologies. Petroleum. 2016;**2**:26-39

[17] Rogala A, Ksiezniak K, Krzysiek J, Hupka J. Carbon dioxide sequestration during shale gas recovery. Physicochemical Problems of Mineral Processing. 2014;**50**

[18] Clark CE, Burnham AJ, Harto CB, Horner RM. Introduction: The Technology and Policy of Hydraulic Fracturing and Potential Environmental Impacts of Shale Gas Development. Taylor & Francis; 2012

[19] Hyne NJ. Nontechnical Guide to Petroleum Geology, Exploration, Drilling, and Production. United States: PennWell Books; 2012

[20] Gandossi L. An overview of hydraulic fracturing and other formation stimulation technologies for shale gas production. The Netherlands: Eur. Commisison Jt. Res. Cent. Tech. Reports. 2013

[21] Barati R, Liang JT. A review of fracturing fluid systems used for hydraulic fracturing of oil and gas wells. Journal of Applied Polymer Science. 2014;**131**

[22] Plan to Study the Potential Impacts of Hydraulic Fracturing on Drinking Water Resources. Washington, DC: US Environmental Protection Agency; 2011

[23] Crews JB, Gomaa AM. Nanoparticle associated surfactant micellar fluids: An alternative to crosslinked polymer systems. In: Presented at the SPE International Oilfield Nanotechnology Conference and Exhibition; Noordwijk, The Netherlands. 2012

[24] Middleton RS, Carey JW, Currier RP, Hyman JD, Kang Q, Karra S, et al. Shale gas and non-aqueous fracturing fluids: Opportunities and challenges for supercritical CO_2. Applied Energy. 2015;**147**:500-509

[25] Kong B, Wang S, Chen S, Dong K. Minimize formation damage in water-sensitive unconventional reservoirs by using energized fracturing fluid. In: SPE International Conference and Exhibition on Formation Damage Control. 2016

[26] Garbis SJ, Taylor JL III. The utility of CO_2 as an energizing component for fracturing fluids. SPE Production Engineering. 1986;**1**:351-358

[27] Gandossi L. State of the Art Report on Waterless Stimulation Techniques for Shale Formations. The Netherlands: Publications Office of the European Union; 2016

[28] Ribeiro L, Sharma M. Fluid selection for energized fracture treatments. In: SPE Hydraulic Fracturing Technology Conference. 2013

[29] Warpinski NR, Mayerhofer MJ, Vincent MC, Cipolla CL, Lolon E. Stimulating unconventional reservoirs: Maximizing network growth while optimizing fracture conductivity. Journal of Canadian Petroleum Technology. 2009;**48**:39-51

[30] McAndrew JJ. Extending the Application of Foam Hydraulic Fracturing Fluids. Colorado, USA: SPE/AAPG/SEG Unconventional Resources Technology Conference; 25-27 August 2014

[31] Parekh B, Sharma MM. Cleanup of water blocks in depleted low-permeability reservoirs. In: SPE Annual Technical Conference and Exhibition. 2004

[32] Gu M. Shale fracturing enhancement by using polymer-free foams and ultra-light weight proppants [PhD thesis]. The University of Texas at Austin; 2013

[33] Oraki Kohshour I, Leshchyshyn T, Munro J, Cassey M, Adejumo A T, Ahmed U, et al. Examination of water management challenges and solutions in shale resource development-could waterless fracturing technologies work?, In Unconventional Resources Technology Conference; 1-3 August 2016; San Antonio, Texas 2016, pp. 2597-2632

[34] Linde. Oil and Gas Reservoirs: Energized Solutions. Germany: The Linde Group; 2014

[35] Liu F, Ellett K, Xiao Y, Rupp JA. Assessing the feasibility of CO_2

storage in the New Albany Shale (Devonian–Mississippian) with potential enhanced gas recovery using reservoir simulation. International Journal of Greenhouse Gas Control. 2013;**17**:111-126

[36] Luo X, Wang S, Wang Z, Jing Z, Lv M. Experimental research on rheological properties and proppant transport performance of GRF–CO_2 fracturing fluid. Journal of Petroleum Science and Engineering. 2014;**120**:154-162

[37] Zelenev AS. Surface energy of north american shales and its role in interaction of shale with surfactants and microemulsions. In: SPE International Symposium on Oilfield Chemistry. 2011

[38] Ahmed S, Elraies KA, Tan IM, Hashmet MR. Experimental investigation of associative polymer performance for CO_2 foam enhanced oil recovery. Journal of Petroleum Science and Engineering. 2017;**157**:971-979

[39] Ahmed S, Elraies KA, Hashmet MR, Hanamertani AS. Viscosity models for polymer free CO_2 foam fracturing fluid with the effect of surfactant concentration, salinity and shear rate. Energies. 2017;**10**

[40] Fei Y, Pokalai K, Johnson R, Gonzalez M, Haghighi M. Experimental and simulation study of foam stability and the effects on hydraulic fracture proppant placement. Journal of Natural Gas Science and Engineering. 2017;**46**:544-554

[41] Sun X, Liang X, Wang S, Lu Y. Experimental study on the rheology of CO_2 viscoelastic surfactant foam fracturing fluid. Journal of Petroleum Science and Engineering. 2014;**119**:104-111

[42] Edrisi A, Kam SI. A new foam model in pipes for drilling and fracturing application. SPE Journal. 2014;**19**:576-585

[43] Edrisi AR, Kam SI. A new foam rheology model for shale-gas foam fracturing applications. In: SPE Canadian Unconventional Resources Conference. 2012

[44] Hirasaki GJ, Lawson JB. Mechanisms of foam flow in porous media: Apparent viscosity in smooth capillaries. Society of Petroleum Engineers Journal. 1985;**25**

[45] Osei-Bonsu K, Shokri N, Grassia P. Fundamental investigation of foam flow in a liquid-filled Hele-Shaw cell. Journal of Colloid and Interface Science. 2016;**462**:288-296

[46] Llave F, Chung F-H, Louvier R, Hudgins D. Foams as mobility control agents for oil recovery by gas displacement. In: Presented at the SPE/DOE Enhanced Oil Recovery Symposium; Tulsa, Oklahoma. 1990

[47] Xiao C, Balasubramanian SN, Clapp LW. Rheology of supercritical CO_2 foam stabilized by nanoparticles. Presented at the SPE Improved Oil Recovery Conference; Tulsa, Oklahoma, USA. 2016

[48] Herzhaft B. Rheology of aqueous foams: A literature review of some experimental works. Oil & Gas Science and Technology. 1999;**54**:587-596

[49] Stevenson P. Foam Engineering: Fundamentals and Applications. United Kingdom: John Wiley & Sons; 2012

[50] Cui L. Application of Foam for Mobility Control in Enhanced Oil Recovery (EOR) Process. United States: Rice University; 2014

[51] Ma K. Transport of surfactant and foam in porous media for enhanced oil recovery processes [PhD thesis]. Texas: Rice University; 2013

[52] Vincent-Bonnieu S, Jones S. Comparative study of foam stability

in bulk and porous media [MSc thesis]. TU Delft: Delft University of Technology; 2014

[53] Ahmed S, Elraies KA, Forooozesh J, Shafian SRBM, Hashmet MR, Hsia ICC, et al. Experimental investigation of immiscible supercritical carbon dioxide foam rheology for improved oil recovery. Journal of Earth Science. 2017;**28**:835-841

[54] Ma K, Ren G, Mateen K, Morel D, Cordelier P. Literature review of modeling techniques for foam flow through porous media. In: Presented at the SPE Improved Oil Recovery Symposium; Tulsa, Oklahoma, USA. 2014

[55] Zhang Z, Freedman VL, Zhong L. Foam Transport in Porous Media—A Review. Washington: Pacific Northwest National Laboratory; 2009

[56] Wang J, Liu H, Ning Z, Zhang H. Experimental research and quantitative characterization of nitrogen foam blocking characteristics. Energy & Fuels. 2012;**26**:5152-5163

[57] Li C, Huang Y, Sun X, Gao R, Zeng FB, Tontiwachwuthikul P, et al. Rheological properties study of foam fracturing fluid using CO_2 and surfactant. Chemical Engineering Science. 2017;**170**:720-730

[58] Farzaneh SA, Sohrabi M. Experimental investigation of CO_2-foam stability improvement by alkaline in the presence of crude oil. Chemical Engineering Research and Design. 2015;**94**:375-389

[59] Sheng J. Enhanced Oil Recovery Field Case Studies. United States: Gulf Professional Publishing; 2013

[60] Kapetas L, Bonnieu SV, Danelis S, Rossen W, Farajzadeh R, Eftekhari A, et al. Effect of temperature on foam flow in porous media. Journal of Industrial and Engineering Chemistry. 2016;**36**:229-237

[61] Langevin D. Influence of interfacial rheology on foam and emulsion properties. Advances in Colloid and Interface Science. 2000;**88**:209-222

[62] Chen Y, Elhag AS, Worthen AJ, Reddy PP, Ou AM, Hirasaki GJ, et al. High temperature CO_2-in-water foams stabilized with cationic quaternary ammonium surfactants. Journal of Chemical & Engineering Data. 2016;**61**:2761-2770

[63] Pramudita RA, Ryoo WS. Viscosity measurements of CO_2-in-water foam with dodecyl polypropoxy sulfate surfactants for enhanced oil recovery application. Korea-Australia Rheology Journal. 2016;**28**:237-241

[64] Wang C, Li HZ. Foam stability of solvent/surfactant/heavy-oil system under reservior conditions. In: SPE International Heavy Oil Conference and Exhibition. 2014

[65] Aronson A, Bergeron V, Fagan ME, Radke C. The influence of disjoining pressure on foam stability and flow in porous media. Colloids and Surfaces A: Physicochemical and Engineering Aspects. 1994;**83**:109-120

[66] Apaydin OG, Kovscek AR. Surfactant concentration and end effects on foam flow in porous media. Transport in Porous Media. 2001;**43**:511-536

[67] Schramm LL. Emulsions, Foams, and Suspensions: Fundamentals and Applications. Germany: John Wiley & Sons; 2006

[68] Wei P, Pu W, Sun L, Wang B. Research on nitrogen foam for enhancing oil recovery in harsh reservoirs. Journal of Petroleum Science and Engineering. 2017;**157**:27-38

[69] Tadros TF. Applied Surfactants: Principles and Applications. Germany: John Wiley & Sons; 2006

[70] Bergeron V. Disjoining pressures and film stability of alkyltrimethylammonium bromide foam films. Langmuir. 1997;**13**:3474-3482

[71] Stubenrauch C, Miller R. Stability of foam films and surface rheology: An oscillating bubble study at low frequencies. The Journal of Physical Chemistry B. 2004;**108**:6412-6421

[72] Espert A, Klitzing RV, Poulin P, Colin A, Zana R, Langevin D. Behavior of soap films stabilized by a cationic dimeric surfactant. Langmuir. 1998;**14**:4251-4260

[73] Georgieva D, Cagna A, Langevin D. Link between surface elasticity and foam stability. Soft Matter. 2009;**5**:2063-2071

[74] Xue Z, Worthen AJ, Da C, Qajar A, Ketchum IR, Alzobaidi S, et al. Ultradry carbon dioxide-in-water foams with viscoelastic aqueous phases. Langmuir. 2015;**32**:28-37

[75] Walker LM. Rheology and structure of worm-like micelles. Current Opinion in Colloid & Interface Science. 2001;**6**:451-456

[76] Parker A, Fieber W. Viscoelasticity of anionic wormlike micelles: Effects of ionic strength and small hydrophobic molecules. Soft Matter. 2013;**9**:1203-1213

[77] Nguyen Q, Hirasaki G, Johnston K. Novel CO_2 Foam Concepts and Injection Schemes for Improving CO_2 Sweep Efficiency in Sandstone and Carbonate Hydrocarbon Formations. Texas, USA: Univ. of Texas at Austin; 2015

[78] Green DW, Willhite GP. Enhanced Oil Recovery: Henry L. Doherty Memorial Fund of AIME. Society of Petroleum Engineers; 1998

[79] Batôt G, Fleury M, Nabzar L. Study of CO_2 Foam Performance in a CCS Context. In: The 30th International Symposium of the Society of Core Analysts-Snowmass. 2016

[80] Sherif T, Ahmed R, Shah S, Amani M. Rheological correlations for oil-based drilling foams. Journal of Natural Gas Science and Engineering. 2016;**35**:1249-1260

[81] Ostwald W. Ueber die Geschwindigkeitsfunktion der Viskosität disperser Systeme. I. Kolloid-Zeitschrift. 1925;**36**:99-117

[82] Abbaszadeh M, Kazemi Nia Korrani A, Lopez-Salinas JL, Rodriguez-de La Garza F, Villavicencio Pino A, Hirasaki G. Experimentally-Based Empirical Foam Modeling. In: Presented at the SPE Improved Oil Recovery Symposium; Tulsa, Oklahoma, USA. 2014

[83] Wanniarachchi W, Ranjith P, Perera M. Shale gas fracturing using foam-based fracturing fluid: A review. Environmental Earth Sciences. 2017;**76**:91

[84] Ahmed S, Elraies K, Hashmet M, Alnarabiji M. Empirical modeling of the viscosity of supercritical carbon dioxide foam fracturing fluid under different downhole conditions. Energies. 2018;**11**:782

[85] Jones S, Laskaris G, Vincent-Bonnieu S, Farajzadeh R, Rossen W. Effect of surfactant concentration on foam: From coreflood experiments to implicit-texture foam-model parameters. Journal of Industrial and Engineering Chemistry. 2016;**37**:268-276

[86] Patton JT, Holbrook ST, Hsu W. Rheology of mobility-control foams.

Society of Petroleum Engineers Journal. 1983;**23**:456-460

[87] Qiang JJ, Li Y, Yang X, Zhang Z, Yao J, Zhu H. Investigation of foam properties and its action on flow improvement of heavy oil. Presented at the International Oil and Gas Conference and Exhibition; Beijing, China. 2010

[88] Bonilla LF, Shah SN. Experimental investigation on the rheology of foams. In: Presented at the SPE/CERI Gas Technology Symposium; Alberta, Canada. 2000

[89] Penny GS, Pursley JT. Field studies of drilling and completion fluids to minimize damage and enhance gas production in unconventional reservoirs. In: European Formation Damage Conference. 2007

[90] Paktinat J, Pinkhouse JA, Stoner WP, Williams C, Carder GA, Penny GS. Case histories: Post-frac fluid recovery improvements of appalachian basin gas reservoirs. In: SPE Eastern Regional Meeting. 2005

[91] Penny GS, Zelenev A, Lett N, Paktinat J, O'Neil BJ. Nano surfactant system improves post frac oil and gas recovery in hydrocarbon rich gas reservoirs. In: SPE Improved Oil Recovery Symposium. 2012

[92] Penny GS, Zelenev AS, Long W, Lett NL, Crafton JW. Laboratory and field evaluation of proppants and surfactants used in fracturing of hydrocarbon rich gas reservoirs. In: SPE Annual Technical Conference and Exhibition. 2012

[93] Lin L, Danican SC, Mueller FA. A novel foaming agent for hydraulic fracturing: Laboratory investigation and field usage. In: SPE Eastern Regional Meeting; West Virginia, USA. 2009

[94] Al-Muntasheri GA, Liang F, Hull KL. Nanoparticle-enhanced hydraulic-fracturing fluids: A review. SPE Production & Operations. 2017;**32**:186-195

[95] Xue Z, Worthen A, Qajar A, Robert I, Bryant SL, Huh C, et al. Viscosity and stability of ultra-high internal phase CO_2-in-water foams stabilized with surfactants and nanoparticles with or without polyelectrolytes. Journal of Colloid and Interface Science. 2016;**461**:383-395

[96] Worthen AJ, Bagaria HG, Chen Y, Bryant SL, Huh C, Johnston KP. Nanoparticle-stabilized carbon dioxide-in-water foams with fine texture. Journal of Colloid and Interface Science. 2013;**391**:142-151

[97] Wilk K, Kasza P, Labus K. Analysis of the applicability of foamed fracturing fluids. Nafta-Gaz. 2015;**71**

[98] Tong S, Singh R, Mohanty KK. A visualization study of proppant transport in foam fracturing fluids. Journal of Natural Gas Science and Engineering. 2018;**52**:235-247

Chapter 3

Thermodynamics of Thermal Diffusion Factors in Hydrocarbon Mixtures

Keshawa Shukla

Abstract

The reliable evaluation of thermal diffusion factors is important to understand the composition variation of the mixture components in hydrocarbon reservoirs. A thermodynamic model of thermal diffusion factors of hydrocarbon mixtures is presented. The model is based on the statistical theory of linear transport of intermolecular forces and accounts for the explicit effects of molecular mass, energy and size parameters. The accuracy of the model is first evaluated by comparing calculated results with the available non-equilibrium molecular dynamics simulation results. The theoretical model is then applied to explain thermal diffusion factors in some selected binary hydrocarbon mixtures over a range of temperature, pressure and molecular composition conditions.

Keywords: hydrocarbon mixtures, thermodynamics, thermal diffusion, molecular interaction, reservoirs

1. Introduction

This chapter deals with one of the key aspects of hydrocarbon production from the oil and gas reservoirs, known as the thermal diffusion process. This process plays an important role to separate isotopic mixtures and isobaric mixtures, analyze hydrodynamic instability in mixtures, transport mass in living matters, migrate minerals, separate and characterize polymers and colloidal particles by thermal field flow fractionation. In the case of hydrocarbon productions, the thermal diffusion process is generally used to study the compositional variation and segregation in hydrocarbon oil and gas reservoirs [1, 2].

In some hydrocarbon reservoirs, very large compositional variations can be observed in horizontal and vertical directions. There is a large temperature gradient in the vertical direction, and a small temperature gradient in the horizontal direction. The horizontal temperature gradient always induces both thermal convection and thermal diffusion, while the vertical temperature gradient causes thermal diffusion but may or may not induce thermal convection. The temperature gradient develops a concentration gradient of the mixture constituents. A thermal diffusion process takes place when the convection free gas and liquid mixture tend to separate under a temperature gradient. This phenomenon is known as the "Soret Effect." This effect can be measured by means of thermal diffusion factor (α_T).

A significant progress has been made in recent past to measure thermal diffusion factors in liquid mixtures [3, 4].

The available experimental data have indicated that in low pressure gaseous mixtures and ideal liquid mixtures, α_T is small, and molecular size and mass govern its magnitude [5]. On the other hand, α_T may be large in non-ideal liquid mixtures where energy interactions, size and shape of the molecules, and thermodynamic conditions govern the magnitude of α_T [6]. Also, the thermal diffusion can enhance composition gradient in vertical direction in hydrocarbon reservoirs [7], and it can enhance or weaken compositional variation in vertical direction as thermal diffusion is usually opposite in sign to the gravitational segregation in hydrocarbons reservoirs [8, 9].

Therefore, a reliable theoretical model of thermal diffusion factors in hydrocarbon mixtures is require to accurately predict the compositional variations in reservoirs and evaluate the formation fluid.

There has been a continued interest in the thermodynamic modeling and the measurement of thermal diffusion factors in multicomponent mixtures (e.g., [10–16]). Numerous classical thermodynamic approaches have been utilized to describe the thermal diffusion factors in binary hydrocarbon mixtures only qualitatively [10–13]. A more successful thermodynamic model of thermal diffusion factors of non-ideal mixtures was presented by Shukla and Firoozabadi [2, 14]. The model was based on the thermodynamics of irreversible processes and kinetic theory, combining both the equilibrium and non-equilibrium properties. In this model, the equilibrium properties were determined from the equation of state while the non-equilibrium properties were obtained from the fluid viscosity. The model predictions of thermal diffusion factors in several binary mixtures were found to represent the experimental data very well. However, the signs of thermal diffusion factors of components could not establish uniquely the direction of warmer and colder region of fluids, especially in multicomponent mixtures. Furthermore, these models provide little or no information about the intermolecular forces in the system, and cannot describe adequately the thermal diffusion factors close to the critical point of the fluids.

Several attempts were made in the past to better describe thermal diffusion factors of gaseous hydrocarbon mixtures using statistical thermodynamics and molecular simulations depending on the intermolecular interactions [17–19]. A review of the developments in the theory and experiment of thermodiffusion has been presented recently by Kohler and Morozov [20]. However, the statistical theory of thermal diffusion was not applied to the liquid hydrocarbon mixtures of industrial interest. Moreover, the rigorous theories were not available to express the volumetric and heat flow properties accurately.

Recently, Shukla [21] proposed a model of thermal diffusion factors in hydrocarbon mixtures using the statistical thermodynamics of intermolecular interactions [22]. The model was able to describe the thermal diffusion factors of several binary hydrocarbon and non-hydrocarbon mixtures. The objective of this paper is to examine the accuracy of the proposed model in describing thermal diffusion factors of binary hydrocarbon mixtures.

Section 2 describes briefly the relation between the mass flux and thermal diffusion coefficient of a binary fluid mixture. Section 3 establishes the relation between statistical thermodynamics of thermal diffusion factors and how to account for the intermolecular interactions of the molecular constituents. Section 4 compares theoretical results with experimental data and examines the reliability of the theory for the selected binary hydrocarbon mixtures. Section 5 presents the conclusion of this study.

2. Thermal diffusion coefficients

We consider a binary fluid mixture. The total diffusive mass flux of component 1 of the mixture is given by [21, 23]

$$\vec{J}_1^{(t)} = -(\rho_m^2/\rho)M_1M_2D_{12}\left[\nabla x_1 + \frac{M_1x_1}{RT}\left(\frac{\overline{V_1}}{M_1} - \frac{1}{\rho}\right)\nabla P/F_1 - \kappa_T\nabla \ln T\right] \quad (1)$$

where, ρ_m is total molar density, M is molecular weight, T is temperature, x_1 is mole fraction of component 1, ρ is mass density, P is pressure, R is gas constant, $\overline{V}_1$ is the partial molar volume of component 1, D_{12} is molecular diffusion coefficient, κ_T is thermal diffusion ratio of component 1, ∇ is the gradient operator, and

$$F_1 = \left(\partial \ln f_1/\partial \ln x_1\right)_{T,P} \quad (2)$$

where f_1 is the fugacity of component 1.

The first, second and third parts of Eq. (1) arise due to the molecular diffusion, pressure diffusion and thermal diffusion. The thermal diffusion factor α_T of component 1 is defined as

$$\alpha_T = \kappa_T/x_1x_2 \quad (3)$$

For a binary mixture, the thermal diffusion factor of component 2 has the opposite sign.

Here we consider a one dimension case in steady state, and assume that there are no convection and gravity segregation. Therefore, the mass flux can be assumed to be zero. Under these conditions, the composition and the temperature gradients are related through the following equation [2]:

$$dx_1/dz = (\alpha_T \ x_1 \ x_2 \ d\ln T/dz) \quad (4)$$

3. Thermodynamics of thermal diffusion factor

Here we present the thermodynamic theory based on the modified form of Chapman and Cowling [22] and Kihara [24] as applied to binary hydrocarbon mixtures. This approach involves the calculation of collision integrals of the fluid mixture for a well-defined potential function. The calculation of the transport property collision integrals for gases, whose molecules obey a simple intermolecular potential, enables to explain the transport properties of slightly non-ideal gas mixtures following the isotropic intermolecular interactions. For non-ideal mixtures, in which molecules interact with strongly anisotropic intermolecular interactions, additional contributions are assumed arising from the expansion of non-equilibrium distributions. These anisotropic interactions could affect the thermal diffusion factors significantly [25].

We consider a binary mixture of components i and j. In this mixture the molecules are assumed to interact with an effective pair-wise additive intermolecular potential function (Exp-6), given by

$$u_{ij}(r) = \varepsilon_{ij}\left[\frac{\alpha_{ij}}{\alpha_{ij}-6}\exp\left(\alpha_{ij}\{1 - r/r_{mij}\}\right) - \frac{\alpha_{ij}}{\alpha_{ij}-6}\left(r_{mij}/r\right)^6\right] \quad (5)$$

where u_{ij} is the potential energy of two molecules of species i and j at a separation distance r, ε_{ij} is the depth of the potential minimum which is located at r_{mij}, α_{ij} determines the softness of the repulsion energy, and k_B is the Boltzmann's constant. In this (Exp-6) potential function, the molecules of mixture species are represented by the size parameter (R_{mab}) and energy parameter (ε_{ab}/k_B).

Using the mth-order Chapman-Cowling approximation, the general form of the thermal diffusion factor (α_T) can be given by

$$(\alpha_T)_m = \frac{5}{2x_1x_2A_{00}^{(m)}}\left[x_1A_{01}^{(m)}\left(\frac{M_1+M_2}{2M_1}\right)^{0.5} + x_2A_{0-1}^{(m)}\left(\frac{M_1+M_2}{2M_2}\right)^{0.5}\right] \tag{6}$$

where, x_1 and x_2 are the mole fractions. M_1 and M_2 are the molecular weights of the mixture components 1 and 2. $A^{(m)}$ is a determinant of (2 m + 1) order, whose general term is A_{ij}, where i and j range from −m to +m, including zero. The minor of $A^{(m)}$ obtained by striking out the row and column containing A_{ij} is denoted by the symbol $A_{ij}^{(m)}$. Similarly, the i and jth minor of $A_{00}^{(m)}$ is denoted by the symbol $A_{ij00}^{(m)}$. The elements A_{ij} are functions of the mole fractions, molecular weights and collision integrals, which are functions of temperature, molecular size and energy parameters.

From Eqs. (3) and (7) the mth-order thermal diffusion ratio $(k_T)_m$ can be defined as

$$(k_T)_m = x_1x_2(\alpha_T)_m \tag{7}$$

and the collision integrals are given by the following equations:

$$\Omega^{(l)}(n) = \left(\frac{k_BT}{2\pi\mu}\right)^{0.5}\int_0^\infty \exp\left(-\gamma^2\right)\gamma^{2n+3}Q^{(l)}(g)d\gamma \tag{8}$$

$$Q^{(l)}(g) = 2\pi\ \left(1 - \cos^l\chi\right)bdb \tag{9}$$

with

$$\gamma^2 = \frac{\mu g^2}{2k_BT} \tag{10}$$

where μ is the reduced mass of a pair of colliding molecules, and g is the initial relative speed of the colliding pair. The molecules are deflected by the collision through a relative angle χ which is a function of g and the collision parameter b.

The dimensionless collision integrals of the above equations can be expressed as follows:

$$\Omega^{(l,n)*} = \left(\frac{4}{\sigma^2(1+n)!}\right)^{-1}\left(1-\frac{1+(-1)^l}{2(1+l)}\right)^{-1}\left(\frac{\mu}{2\pi k_BT}\right)^{0.5}\Omega^{(l)}(n) \tag{11}$$

where $\Omega^{(l,n)*}$ is the dimensionless collision integral reduced with respect to that of the diameter σ of a rigid elastic sphere.

Using the first-order approximation, Eq. (7) is written as

$$(\alpha_T)_1 = \left(6C_{12}^* - 5\right)\left[\frac{x_1S_1 - x_2S_2}{x_1^2Q_1 + x_2^2Q_2 + x_1x^{12}Q_{12}}\right] \tag{12}$$

where the parameters S1, S2, Q1 and Q2 are given by

$$S_1 = \frac{M_1}{M_2}\left(\frac{2M_2}{M_1+M_2}\right)^{0.5}\left(\frac{\Omega_{11}^{(2,2)*}}{\Omega_{12}^{(1,1)*}}\right)\left(\frac{\sigma_{11}}{\sigma_{12}}\right)^2 - \frac{4M_1M_2A_{12}^*}{(M_1+M_2)^2} - \frac{15M_2(M_2-M_1)}{2(M_1+M_2)^2} \tag{13}$$

$$S_2 = \frac{M_2}{M_1}\left(\frac{2M_1}{M_1+M_2}\right)^{0.5}\left(\frac{\Omega_{22}^{(2,2)*}}{\Omega_{12}^{(1,1)*}}\right)\left(\frac{\sigma_{22}}{\sigma_{12}}\right)^2 - \frac{4M_1M_2A_{12}^*}{(M_1+M_2)^2} - \frac{15M_1(M_1-M_2)}{2(M_1+M_2)^2} \tag{14}$$

$$Q_1 = \left(\frac{2}{M_2(M_1+M_2)}\right)\left(\frac{2M_2}{(M_1+M_2)}\right)^{0.5}\left(\frac{\Omega_{11}^{(2,2)*}}{\Omega_{12}^{(1,1)*}}\right)\left(\frac{\sigma_{11}}{\sigma_{12}}\right)^2 \left[\left(\frac{5}{2}-\frac{6}{5}B_{12}^*\right)M_1^2 + 3M_2^2 + \frac{8}{5}M_1M_2A_{12}^*\right] \tag{15}$$

$$Q_2 = \left(\frac{2}{M_1(M_1+M_2)}\right)\left(\frac{2M_1}{(M_1+M_2)}\right)^{0.5}\left(\frac{\Omega_{22}^{(2,2)*}}{\Omega_{12}^{(1,1)*}}\right)\left(\frac{\sigma_{22}}{\sigma_{12}}\right)^2 \left[\left(\frac{5}{2}-\frac{6}{5}B_{12}^*\right)M_2^2 + 3M_1^2 + \frac{8}{5}M_1M_2A_{12}^*\right] \tag{16}$$

$$Q_{12} = 15\left(\frac{M_1-M_2}{M_1+M_2}\right)^2\left(\frac{5}{2}-\frac{6}{5}B_{12}^*\right) + \left(\frac{4M_1M_2A_{12}^*}{(M_1+M_2)^2}\right)\left(11-\frac{12}{5}B_{12}^*\right) + \frac{8(M_1+M_2)}{5\left(M_1M^{12}\right)^{0.5}}\left(\frac{\Omega_{11}^{(2,2)*}}{\Omega_{12}^{(1,1)*}}\right)\left(\frac{\Omega_{22}^{(2,2)*}}{\Omega_{12}^{(1,1)*}}\right)\left(\frac{\sigma_{11}}{\sigma_{12}}\right)^2\left(\frac{\sigma_{22}}{\sigma_{12}}\right)^2 \tag{17}$$

A_{12}^*, B_{12}^* and C_{12}^* are functions of the collision integrals as given by

$$A_{12}^* = \left(\frac{\Omega_{12}^{(2,2)*}}{\Omega_{12}^{(1,1)*}}\right) \tag{18}$$

$$B_{12}^* = \left(\frac{5\Omega_{12}^{(1,2)*} - 4\Omega_{12}^{(1,3)*}}{\Omega_{12}^{(1,1)*}}\right) \tag{19}$$

$$C_{12}^* = \left(\frac{\Omega_{12}^{(1,2)*}}{\Omega_{12}^{(1,1)*}}\right) \tag{20}$$

The details on the various parameters are given elsewhere [21].

To consider the effects of pressure and unlike interaction parameters, Eq. (21) for C_{12}^* was modified empirically as follows:

$$C_{12}^* = \left(\frac{\Omega_{12}^{(1,2)*}}{\Omega_{12}^{(1,1)*}}\right)\left[\exp\left(p_1^* f_1 + p_2^* f_2 + p_3^* f_3\right)\right] \tag{21}$$

where p_i are the mixture parameters and f_i terms are given by

$$\begin{aligned} f_1 &= x_1 P_x^*/T_x^* \\ f_2 &= f_1^{\,2} \\ f_3 &= f_1^{\,3} \end{aligned} \tag{22}$$

where the reduced pressure and temperature are given in terms of energy and size parameters of the molecules

$$P_x^* = PR_{mx}^3/\varepsilon_x \quad T_x^* = kT/\varepsilon_x \tag{23}$$

The following van der Waals mixing rules were applied to determine the mixture properties:

$$R_{mx}^3 = \sum_{i,j=1}^{2} x_i x_j R_{mij}^3 \tag{24}$$

$$\varepsilon_x R_{mx}^3 = \sum_{i,j=1}^{2} x_i x_j \varepsilon_{ij} R_{mij}^3 \tag{25}$$

$$R_{mij} = \left(R_{mii} + R_{mjj}\right)/2 \tag{26}$$

$$\varepsilon_{ij} = \left(\varepsilon_{ii}\varepsilon_{jj}\right)^{0.5} \tag{27}$$

Since the properties of the reservoir fluids depend on the fluid compositions, temperature and pressure, and since the collision integrals do not account for the pressure effects of the liquid mixtures, the empirical Eq. (22) was applied to explain the properties of the reservoir fluid mixtures under both temperature and pressure as needed.

4. Results and discussion

In this section, the performance of the thermodynamic theory is examined to represent the thermal diffusion factors in a few selected binary gas and liquid mixtures.

4.1 Potential parameters

The intermolecular potential parameters along with molecular weight of the pure fluids involved in binary mixtures studied in this work are given in **Table 1**. Molecular parameters (ε/k_B, R_m and α) are based on the correlation of the pure fluid viscosity and/or second virial coefficient data at the ambient pressure [25]. The unlike interaction parameters (R_{mij}, ε_{ij}) were evaluated following the arithmetic rule for R_{mij} and geometric rule for ε_{ij}. α was kept the same for all the fluids. The

Fluid	ε/k, K	R_m, A°	α	M
C_1	148.20	3.850	14	16
C_3	281.86	5.331	14	44
C_4	339.35	5.755	14	58
C_5	374.78	6.102	14	72.15
C_7	431.13	6.861	14	100
C_{10}	492.89	7.667	14	142.29
C_{16}	575.42	8.827	14	226

Table 1.
(Exp-6) Potential parameters for the mixture components.

Mixture	P_1	P_2	P_3
$C_1 + C_3$	−51.9	820.6	−2742.7
$C_1 + C_4$	13.2	1.34	−183.7
$C_5 + C_{10}$	−0.6864	0	0
$C_7 + C_{16}$	525.8	−21375.0	294494.2

Table 2.
Polynomial parameters of $C^*(1,2)$.

pressure and potential parameters dependences of the unlike collision integrals are expressed in terms of a simple polynomial as given in Eq. (22) above. The polynomial parameters p_i for the specific mixtures are given in **Table 2**. In this work, no effort was made to optimize the unlike energy, size and α parameters.

4.2 Application of theory to liquid mixtures of hydrocarbons

In order to check the reliability of the thermodynamic model, theoretical results are first tested against the non-equilibrium molecular dynamics simulation results using the same (Exp-6) intermolecular potential function for a binary mixture of pentane and decane ($C_5 + C_{10}$) [26]. Note that the optimized parameters were used in the non-equilibrium molecular dynamics simulations to represent the experimental data of thermal diffusion factors of ($C_5 + C_{10}$) mixture.

Figure 1 compares the theoretical results with the simulation results of α_T for the mixture of pentane and decane ($C_5 + C_{10}$) at 300 K and 0.1 Mpa [26]. These results show that the thermodynamic model can describe α_T reasonably well with the uncertainties of simulation results. Also included in the **Figure 1** are measured data for the system ($C_5 + C_{10}$). The model is seen to compare well with both simulation and measured results within their data uncertainties.

The above results suggest that the thermodynamic model with the pressure dependent collision integrals offers reliable prediction of α_T as a function of both temperature and concentration in binary mixtures. Therefore, we adopted the unlike potential parameter and pressure dependent collision integrals to first

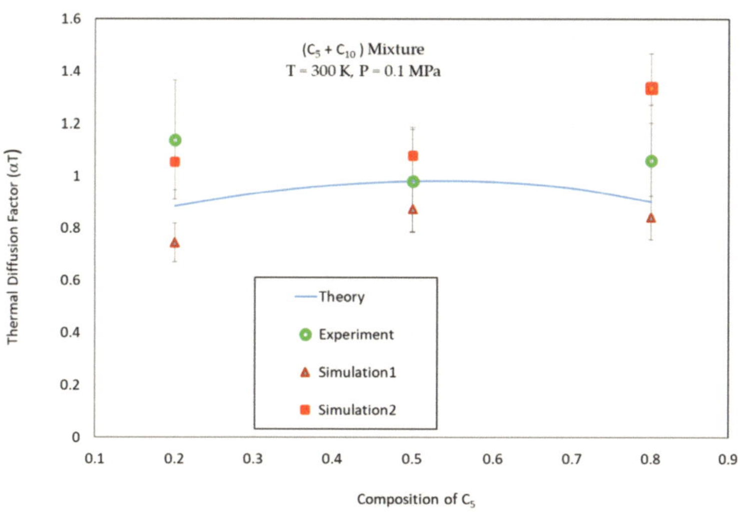

Figure 1.
Thermal diffusion factor for mixture ($C_5 + C_{10}$) from theory, simulation and experiment.

correlate a value of α_T in liquid hydrocarbon mixtures at a single temperature and equimolar condition, and extended that to all other conditions. When several data points of thermal diffusion factor were available in different non-ideal conditions of temperature, pressure and concentration, we re-evaluated the single point parameters by incorporating several data points in parameters regression. Also, in most of the cases the three parameters p_1, p_2 and p_3 can describe well the diffusion factors and can depend on the temperature, pressure, concentration and interaction parameters.

Figure 2 presents results for the collision integrals independent of pressure. The calculated collision integral results are physically consistent and agree with the literature data very well [25].

Figure 3 shows theoretical predictions and experimental data [27] of α_T for mixture (C_1 + C_3) at a given temperature T = 346 K and P = 5.5 Mpa as a function of composition of C_1. **Figure 4** presents theoretical and experimental results of α_T for T = 346 K, and composition, $x_1(C_1)$ = 0.34 as the function of pressure. The latter

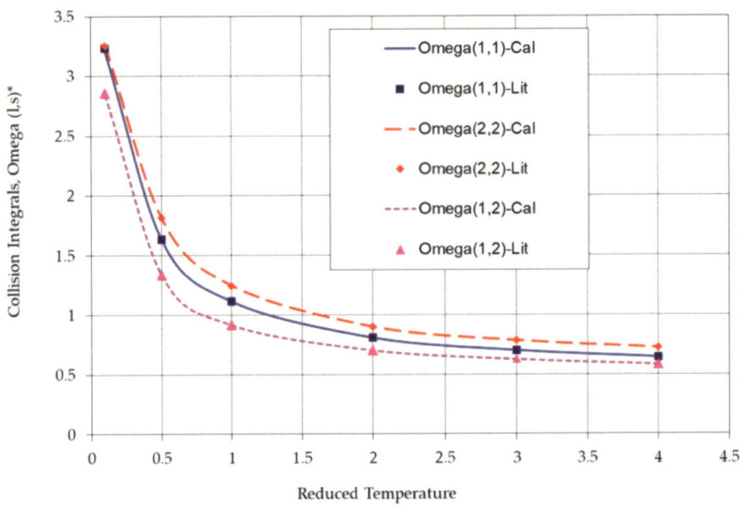

Figure 2.
Collision integrals from calculation and literature.

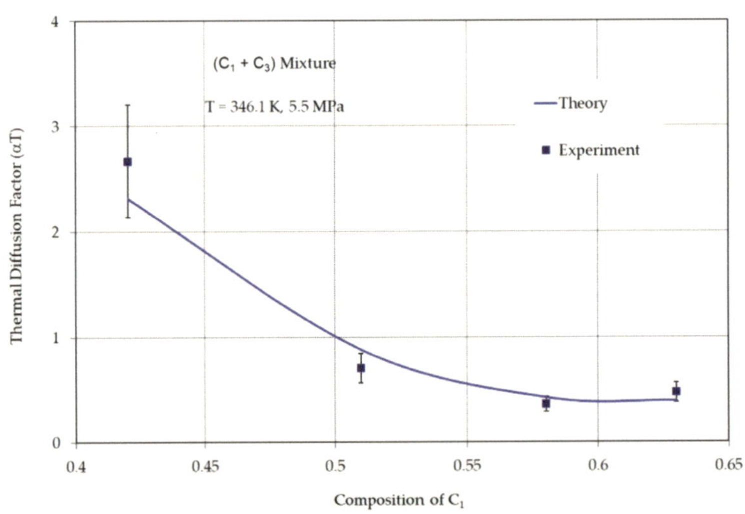

Figure 3.
Thermal diffusion factor for mixture (C_1 + C_3) from theory and experiment.

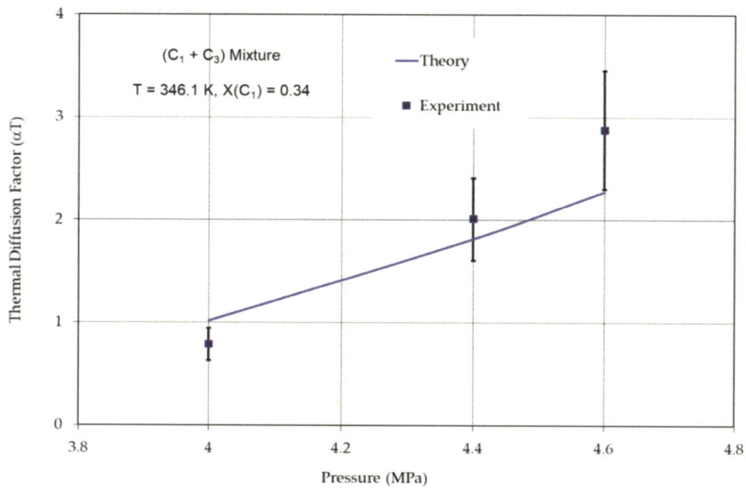

Figure 4.
Thermal diffusion factor for mixture (C_1 + C_3) from theory and experiment.

condition is close to the critical point of the mixture (C_1 + C_3). In both cases the theoretical results are in agreement with the measured data well within the experimental uncertainty.

Figure 5 compares theoretical predictions with experimental data [23] of α_T for mixture of methane and butane (C_1 + C_4) for temperature 346 K and composition $x_1(C_1)$ = 0.34. **Figure 6** presents similar comparisons at the lower temperature of 319 K and composition of $x_1(C_1)$ = 0.49. The variations of thermal diffusion coefficients with pressure are investigated. The comparison between theory and experiment is very good for all the tested conditions.

To further examine the reliability of our models, **Figure 7** compares theoretical and experimental results of α_T for the more non-ideal mixture of Heptane and Hexadecane (C_7 + C_{16}) [28]. The model can describe α_T reasonably well over the whole range of the composition.

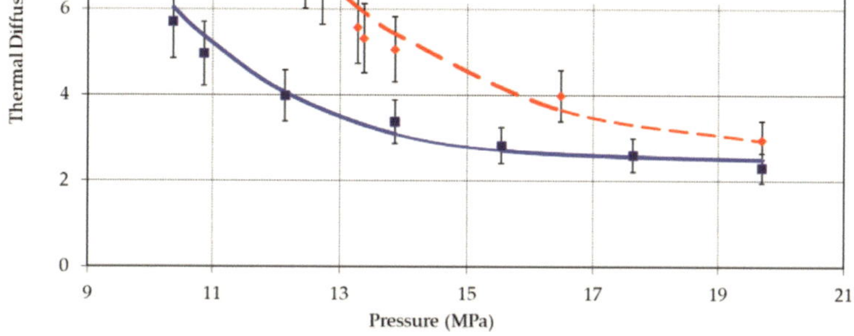

Figure 5.
Thermal diffusion factor for mixture (C_1 + C_4) from theory and experiment.

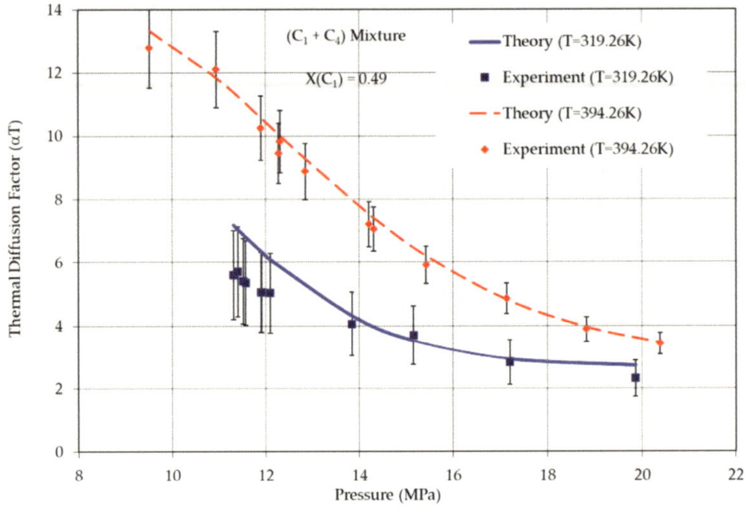

Figure 6.
Thermal diffusion factor for mixture (C_1 + C_4) from theory and experiment.

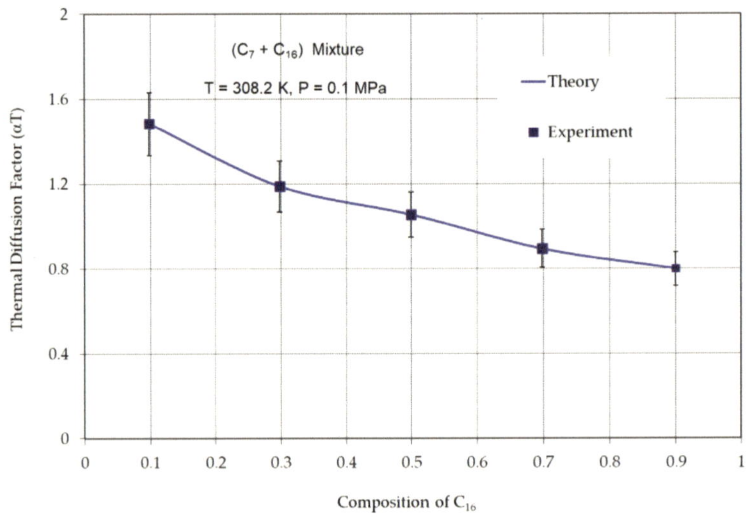

Figure 7.
Thermal diffusion factor for mixture (C_7 + C_{16}) from theory and experiment.

5. Conclusions

In this paper the statistical thermodynamics has been applied to predict the thermal diffusion factors of binary hydrocarbon systems using the thermodynamic model based on the statistical thermodynamics and the (Exp-6) potential function of two-body molecular interactions. The collision integrals were redefined to account for the energy and size parameters of the molecules in addition to their pressure and temperature dependency. Theoretical results are tested against the molecular simulation results and experimental data for a few selected binary hydrocarbon gas and liquid mixtures. The model can successfully describe the simulation results of the binary hydrocarbon mixture investigate her. In general, the comparisons of theoretical results with experimental data for the thermal diffusion coefficients show a very good performance of the theory in different non-ideal reservoir conditions over a range of temperature, pressure and concentration. The unlike interaction parameters are seen to be important for accounting the non-ideal effects in collision integrals, and for improving the correlation and prediction of thermal diffusion factors in non-ideal hydrocarbon liquid mixtures.

Acknowledgements

The author would like to thank the Subsea Engineering Program at College of Engineering, Texas A&M University for investing his time and effort to contribute this chapter to this book on the special topic of "Hydrocarbons and Cross Coupling." The open access publishing fees for this article have been covered by the Texas A&M University Open Access to Knowledge Fund (OAKFund), supported by the University Libraries and the Office of the Vice President for Research.

Conflict of interest

No potential conflict of interest.

Author details

Keshawa Shukla
Subsea Engineering, College of Engineering, Texas A&M University, College Station, TX, USA

*Address all correspondence to: kpshukla@tamu.edu

References

[1] Cussler E. Diffusion: Mass Transfer in Fluid Systems. 2nd ed. Chichester: Cambridge University Press; 1997

[2] Shukla K, Firoozabadi A. Mint: A new model of thermal diffusion coefficients in binary hydrocarbon mixtures. IECR. 1998;**37**:3331-3342

[3] De Mezquia D, Wang Z, Lapeira E, Klein M, Wiegand S, Mint B-AM. Themodiffusion, molecular diffusion and Soret coefficient of binary and ternary mixtures of n-hexane, n-dodecane and toluene. EPJE. 2014;**37**:106

[4] Bou-Ali M, Ahadi A, de Mezquia D, Galand Q, Gebhardt M, Khlybov O, Kohler W, Larranaga M, Legros J, Lyubimova T, Mialdum A, Ryzhkov I, Saghir M, Shevtsova V, Vaerenbergh S. Mint: Benchmark values for the soret, thermodiffusion and molecular diffusion coefficients of the ternary mixture tetralin+isobutylbenzene +n-dodecane with 0.8-0.1-0.1 mass fraction. EPJ. 2015;**38**:30

[5] Vergaftik N. Tables on the Thermophysical Properties of Liquids and Gases. Chichester: Hemisphere; 1975

[6] Kincaid J, Cohen E, Lopez de Haro M. Mint: The Enskog theory for multicomponent mixtures. IV. Thermal diffusion. JCP. 1987;**86**:963

[7] Firoozabadi A, Dindoruk B, Chang E. Mint: Areal and vertical composition variation in hydrocarbon reservoirs: Formulation and binary results. Entropie. 1996;**198/199**:109

[8] Ghorayeb K, Firoozabadi A, Anraku T. Mint: Interpretation of the unusual fluid distribution in the Yufutsa ngas condensate field. SPEJ. 2003;**8**:114

[9] Galliero G, Bataller H, Croccolo F, Vermorel R, Artola P, Rousseau B, Vesovic V, Bou-Ali M, de Z'arate J, Xu S, Zhang K, Mint MF. Impact of thermodiffusion on the initial vertical distribution of species in hydrocarbon reservoirs. Microgravity Science and Technology. 2016;**28**:79

[10] Artola P, Rousseau B, Galliero G. Mint: A new model for thermal diffusion: Kinetic approach. Journal of the American Chemical Society. 2008; **130**:10963

[11] Haase R, Ducker K, Buchner H, Schwinum J. Mint: Thermal diffusion in the critical region of binary liquid system. Zeitschrift für Physikalische Chemie. 1994;**186**:113

[12] Kempers L. Mint: A thermodynamic theory of soret effect in a multicomponent liquid. JCP. 1989;**90**: 6541

[13] Galliero G. Mint: Thermal diffusion in Lennard-Jones fluids in the frame of the law of the corresponding state. FPE. 2004;**224**:13

[14] Firoozabadi A, Ghorayeb K, Shukla K. Mint: Theoretical model of thermal diffusion factors in multicomponent mixtures. AIChEJ. 2000;**46**:892-900

[15] Gallerio G, Montel F. Mint: Understanding compositional grading in petroleum reservoir thanks to molecular simulations. SPE. 2009;**121902**:1

[16] Sharipov F, Benites V. Mint: Transport coefficients of helium-argon mixture based on ab initio potential. JCP. 2015;**143**:154104

[17] Galliero G, Boned C. Mint: Molecular dynamics study of the repulsive form influence of the interaction potential on structural, thermodynamic, interfacial and transport properties. JCP. 2008;**129**: 074506

[18] Artola P, Rousseau B. Mint: Thermal diffusion in simple liquid mixtures: What we have learnt from molecular dynamics simulations? MP. 2013;**111**: 3394

[19] Artola P-A, Rousseau B. Mint: Isotopic soret effect in ternary mixtures: Theoretical predictions and molecular simulations. JCP. 2015;**143**:174503

[20] Kohler W, Morozov K. Mint: The Soret effect in liquid mixtures—A review. Journal of Non-Equilibrium Thermodynamics. 2016;**41**:151

[21] Shukla K. Mint: Statistical thermodynamics of thermal diffusion factors in binary hydrocarbon mixtures —An application. MP. 2017;**115**: 1253-1263

[22] Chapman S, Cowling T. Mint: The Mathematical Theory of Non-uniform Gases. Chichester: Cambridge University Press; 1952

[23] Rutherford W, Roof J. Mint: Thermal diffusion in methane-n-butane mixtures in the critical region. JPC Journal of Planar Chromatography Modern TLC. 1959;**63**:1506

[24] Kihara T. Mint: Imperfect gases. Reviews of Modern Physics. 1953;**25**:831

[25] Maitland G, Rigby M, Smith E, Wakeham W. Intermolecular Forces—Their Origin and Determination. Chichester: Clarendon Press; 1981

[26] Perronace A, Leppla C, Leroy F, Rousseau F, Wiegand S. Mint: Soret and mass diffusion measurements and molecular dynamics simulations of n-pentane-n-decane mixtures. JCP. 2002;**116**:3718

[27] Haase R, Borgmann H, Ducker K, Lee W. Mint: Thermal diffusion in the critical region of binary liquid systems. Zeitschrift für Naturforschung. 1971; **A26**:1224

[28] Shieh J. Mint: Thermal diffusion and segmental motion in binary n-alkane systems. JPC Journal of Planar Chromatography Modern TLC. 1969;**73**: 1508

Section 3

Well and Reservoir Analysis and Simulation

Chapter 4

Well Test Analysis for Hydraulically-Fractured Wells

Freddy Humberto Escobar

Abstract

This chapter focuses on the application of Tiab's direct synthesis (*TDS*) technique for practical and accurate interpretation of pressure tests on vertical wells in conventional reservoirs, so bilinear, linear, and elliptical flow regimes can be used for fracture characterization. Most fractured well interpretation tests are conducted using nonlinear regression analysis if the pressure model is available. This method has some drawbacks associated with the nonuniqueness of the solution. Also, the conventional straight-line method requires one plot for each individual flow regime observed in the pressure tests, and the estimated parameters cannot be verified. Tiab's direct synthesis (*TDS*) methodology, which uses specific lines and intersection points found on the pressure and pressure derivative plot, is used in some direct equations which are obtained from the solution of the diffusivity equation for a given flow regime. It has been proven to provide accurate results, and its power allows verification of most results which is not possible from any other technique. The methodology has been successfully explained and tested by its application in two examples, although there exists more than a hundred articles that provide many useful applications.

Keywords: bilinear flow, linear flow, elliptical flow, half-length fracture, fracture conductivity, hydraulic fracturing

1. Introduction

Throughout their history, well test analyses for fractured wells have received many contributions. For practical purposes, let us name the most important ones for this chapter. A good place to start is by mentioning the work in [1], which described the pressure behavior for infinite-conductivity and uniform-flux fractured wells, so people started conducting interpretation tests on such wells by using type-curve matching. Later, [2] introduced the concept of finite-conductivity fractures and established the onset value of dimensionless conductivity as 300. Values lower than that are considered finite-conductivity values, and those above 300 are classified as infinite conductivity. In [2], a fine semi-analytical solution was introduced for describing the well-pressure behavior in hydraulically fractured wells. This solution was then applied in [3] to provide a well interpretation method using type-curve matching. Since then, other mathematical solutions have been presented for finite-conductivity fractures. Among them, the work in [4] using fractal theory is worth mentioning.

The way of conducting well test interpretation was changed by the introduction of Tiab's direct synthesis (TDS) technique by [5]. This revolutionary and modern technique focuses on the different flow regimes seen on the pressure derivative curve. Defined lines are drawn through each individual flow regime, and the intersection points found among them are read and used for reservoir characterization. Additionally, reading arbitrary points on the pressure and pressure derivative of each flow regime also serve for reservoir parameter determination. A great number of applications of the *TDS* technique are given in [6]. The second work [7], by the same author of [5], presented *TDS* technique for infinite-conductivity and uniform-flux fractures in vertical wells. In [7], the elliptical or biradial flow regime was introduced and characterized. This elliptical flow is also seen in horizontal wells and was characterized in [8–10]. Because of the similarity between the mathematical models of hydraulic fractures and horizontal wells, this concept was applied by [11] to determine the average reservoir pressure in formations drained by horizontal wells using the *TDS* technique. The infinite-conductivity model in [7] also included the late-time pseudosteady-state period as well as some equations involved in the drainage area (conventional analysis for this case was included in [12]). This may be disadvantageous for inexperienced users of *TDS* technique when interpreting pressure tests without reaching reservoir boundaries because the equations involved the use of the unknown reservoir drainage area, although it can be still applied by using the intersection points. To overcome this drawback, [13] presented a new mathematical model excluding the late-time pseudosteady-state period.

TDS technique for finite-conductivity fractured wells is given in [14], with practical field applications to demonstrate the usefulness of the technique. The fracture parameters can be readily obtained by using an arbitrary point on the flow regimes. *TDS* technique plays an important role when analyzing short pressure tests because a user can “make up” nonexisting flow regimes since, for instance, the radial flow horizontal line can be obtained from the reservoir permeability even though radial flow regime is absent. [15, 16] extended the works of finite- and infinite-conductivity fractures in naturally fractured reservoirs. The equations provided by these works can also be applied to either homogeneous or naturally fractured formations since they involve a dummy variable that takes the value of one for the homogeneous case or the value of the dimensionless storativity coefficient for the case of a naturally fractured formation.

TDS technique has also been extended to several scenarios related to hydraulically fractured wells. For instance, when a finite-conductivity fracture intersects with a fault, the pressure trace changes; then, the equations developed in [17] apply for this case. There are cases where a threshold pressure is required to start the flow. The work in [18] includes this concept in uniform-fractured vertical wells, and the work in [19] includes the concept for horizontal wells. Also, when the fractured face is damaged, a pseudolinear flow regime develops along the fracture. [1] included *TDS* technique to characterize such systems. [16] presented *TDS* technique for fractured wells in gas composite reservoirs. *TDS* technique can also be usefully applied to transient-rate analysis, as seen in [20]. Application of *TDS* technique to horizontally isolated fractured wells was presented and characterized in [21] and in conventional analysis in [22]. The works in [23, 24] use *TDS* technique for shale reservoirs. Other applications of *TDS* technique to these systems are given by [25] under transient-rate analysis and [26] for pressure-transient analysis conditions. Other important applications of TDS Technique to fractured wells are given by [29, 30].

This chapter is devoted to the application of *TDS* technique to hydraulically fractured wells in either homogeneous or naturally fractured formations. Without given detailed derivations, the expressions for characterizing the hydraulic fracture parameters are presented along with the way they should be used. Important relationships and practical exercises are included.

2. *TDS* basis

The pioneer publication on the *TDS* technique, [5], explains in detail the derivation of the equations. The Laplace space solution of the arithmetic pressure derivative for a homogeneous and infinite reservoir with skin and wellbore storage is also presented in [5] and given by

$$P_D{}' = \frac{4}{\pi^2}\int_0^\infty \left(\frac{e^{-u^2 t_D}}{u\left\{ \left[uC_D J_0(u) - (1 - C_D s u^2) J_1(u)\right]^2 + \left[uC_D Y_0(u) - (1 - C_D s u^2) Y_1(u)\right]^2 \right\}} \right) du. \tag{1}$$

However, we know that the pressure derivative is a horizontal line during radial flow regime. The dimensionless pressure derivative during radial line is easier represented by

$$t_D{}^* P_D{}' = 0.5. \tag{2}$$

Then, to obtain practical equations, dimensionless parameters must be used. The dimensionless time, based upon half-fracture length and reservoir drainage area, is given below:

$$t_{Dxf} = \frac{0.000263kt}{\phi \mu c_t x_f^2} \tag{3}$$

and

$$t_{DA} = \frac{0.000263kt}{\phi \mu c_t A}. \tag{4}$$

The dimensionless pressure and pressure derivative parameters for oil reservoirs are given by

$$P_D = \frac{kh\Delta P}{141.2q\mu B} \tag{5}$$

and

$$t_D{}^* P_D{}' = \frac{kh(t^* \Delta P')}{141.2q\mu B}. \tag{6}$$

Finally, the dimensionless fracture conductivity introduced in [3] is defined as

$$C_{fD} = \frac{k_f w_f}{k\ x_f}. \tag{7}$$

It is observed from Eq. (5) that the two key parameters of a hydraulic fracture are the half-fracture length, x_f, and the fracture conductivity, $k_f w_f$. The total length of the fracture is given by 2 x_f.

The easiest application of *TDS* technique is given by replacing the dimensionless pressure derivative defined by Eqs. (6) and (2), to provide an expression to readily determine formation permeability:

$$k = \frac{70.6q\mu B}{h(t * \Delta P')_R}, \tag{8}$$

where $(t^*\Delta P')_R$ is the pressure derivative value during radial flow regime. The equations for the *TDS* technique are derived in the same manner Eq. (8) was obtained.

3. Biradial flow regime

Biradial or elliptical flow normally results in a hydraulically fractured well when areal anisotropy is present. This is recognized on the pressure derivative versus time log-log plot by a straight line with a slope of 0.36. In hydraulic fractures, the flow from the formation to the fracture is described by parallel flow lines resulting in a linear flow geometry better known as linear flow regime and characterized by a slope of 1/2 on the pressure derivative versus time log-log plot.

Both linear flow and biradial/elliptical flow regimes are seen on the plot of dimensionless pressure and pressure derivative versus dimensionless time based on half-fracture length for a naturally fractured formation. New expressions for the elliptical flow regime introduced in [13] excluding reservoir drainage area are given by.

$$P_D = \frac{25}{9}\left(\frac{\pi t_{Dxf}}{26\xi}\right)^{0.36} \tag{9}$$

and

$$t_D{}^*P_D{}' = \left(\frac{\pi t_{Dxf}}{26\xi}\right)^{0.36}, \tag{10}$$

being ξ a dummy variable that defines either a homogeneous or naturally fractured formation. When $\xi = 1$, a homogeneous reservoir is considered. For the case of naturally fractured formations, $\xi = \omega$, the dimensionless storativity coefficient.

Once dimensionless parameters given by Eqs. (3), (5), and (6) are replaced into Eqs. (9) and (10), respectively, and solve for the half-fracture length, which yields

$$x_f = 22.5632\left(\frac{qB}{h(\Delta P)_{BR}}\right)^{1.3889}\sqrt{\frac{t_{BR}}{\xi\phi c_t}\left(\frac{\mu}{k}\right)^{1.778}} \tag{11}$$

and

$$x_f = 5.4595\left(\frac{qB}{h(t*\Delta P')_{BR}}\right)^{1.3889}\sqrt{\frac{t_{BR}}{\xi\phi c_t}\left(\frac{\mu}{k}\right)^{1.778}}. \tag{12}$$

TDS technique is based on drawing a straight line throughout a given flow regime; then, the user is expected to read the pressure, ΔP_{BR}, and pressure derivative, $(t^*\Delta P')_{BR}$, at a given time, t_{BR}. A better way to reduce noise effects consists of extrapolating the mentioned straight line (biradial for this case) to the time of 1 h and read the pressure derivative value, $(t^*\Delta P')_{BR1}$, at 1 h. For this case, the pressure and pressure derivative set in Eqs. (11) and (12) is changed to ΔP_{BR1} and $(t^*\Delta P')_{BR1}$, respectively.

When bilinear flow is unseen, fracture conductivity can be found with an expression presented in [27]

$$k_f w_f = \frac{3.31739k}{\frac{e^s}{r_w} - \frac{1.92173}{x_f}}. \tag{13}$$

[5] also provided an equation for the determination of the skin factor using an arbitrary point read during radial flow regime:

$$s = 0.5\left\{\frac{\Delta P_R}{(t^*\Delta P')_R} - \ln\left(\frac{kt_R}{\phi\mu c_t r_w^2}\right) + 7.43\right\}. \tag{14}$$

The pseudosteady-state regime governing the pressure derivative equation is given by

$$[t_{DA}{}^*P_D{}']_P = 2\pi(t_{DA})_P. \tag{15}$$

[7] used the point of intersection, t_{RPi}, of Eqs. (2) and (15) to derive an equation for the estimation of the drainage area:

$$A = \frac{kt_{RPi}}{301.77\phi\mu c_t}. \tag{16}$$

The derivation of Eq. (16) follows a similar idea as that presented later in Section 4 for the use of the points of intersection.

4. Bilinear and linear flow regimes

Bilinear flow regime takes place when two linear flows, normal one flowing into the other, take place simultaneously. This situation occurs in low conductivity fractures where linear flow along the fracture and linear flow from the formation to the fracture are observed. Bilinear flow is recognized in the pressure derivative curve by a slope of 0.25. However, this is not shown in **Figure 1** since bilinear flow is absent. The governing expressions for early bilinear and linear flow regimes for vertical fractures in naturally fractured systems were, respectively, presented in [16]

$$P_D = \frac{2.45}{\sqrt{C_{fD}}}\left(\frac{t_{Dxf}}{\xi}\right)^{1/4}, \tag{17}$$

$$t_D{}^*P_D{}' = \frac{0.6125}{\sqrt{C_{fD}}}\left(\frac{t_{Dxf}}{\xi}\right)^{1/4}, \tag{18}$$

$$P_D = \left(\frac{\pi t_{Dxf}}{\xi}\right)^{1/2}, \tag{19}$$

and

$$t_D{}^*P_D{}' = \frac{1}{2}\left(\frac{\pi t_{Dxf}}{\xi}\right)^{1/2}. \tag{20}$$

Linear flow regime can be used to find the half-fracture length, and bilinear flow regime allows finding the fracture conductivity. Once the dimensionless quantities of Eqs. (1) and (3)–(5) are replaced in Eqs. (16)–(19), the fracture conductivity is solved for then

$$k_f w_f = \frac{1947.46}{\sqrt{\xi\phi\mu c_t k}}\left(\frac{q\mu B}{h(\Delta P)_{BL1}}\right)^2, \tag{21}$$

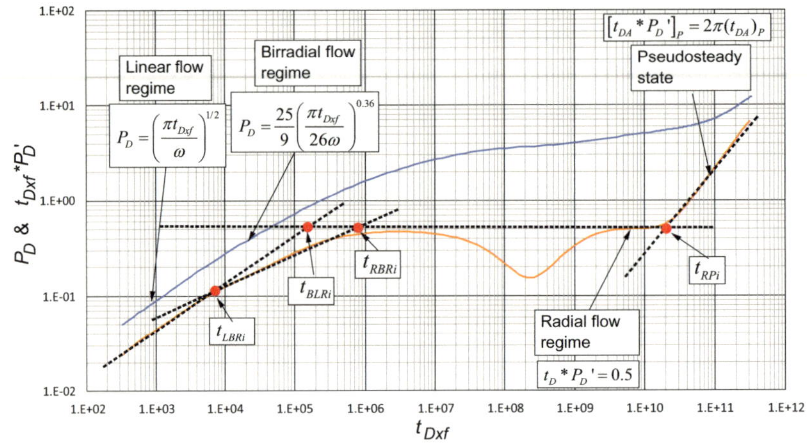

Figure 1.
Dimensionless pressure and pressure derivative behavior for an infinite-conductivity fractured vertical well in a naturally fractured bounded reservoir, λ = 1 × 10^{-9} and ω = 0.1 (taken from [13]).

$$k_f w_f = \frac{121.74}{\sqrt{\xi \phi \mu c_t k}} \left(\frac{q \mu B}{h(t * \Delta P')_{BL1}} \right)^2, \tag{22}$$

Once the fracture conductivity is found, Eq. (7) applies to find the dimensionless fracture conductivity if reservoir permeability and the half-fracture length are known. When bilinear flow is absent, the fracture conductivity may be found from Eq. (13), or the dimensionless fracture conductivity can be read from **Figure 2**:

$$x_f = \frac{4.064 q B}{h(\Delta P')_{L1}} \sqrt{\frac{\mu}{\xi \phi c_t k}} \tag{23}$$

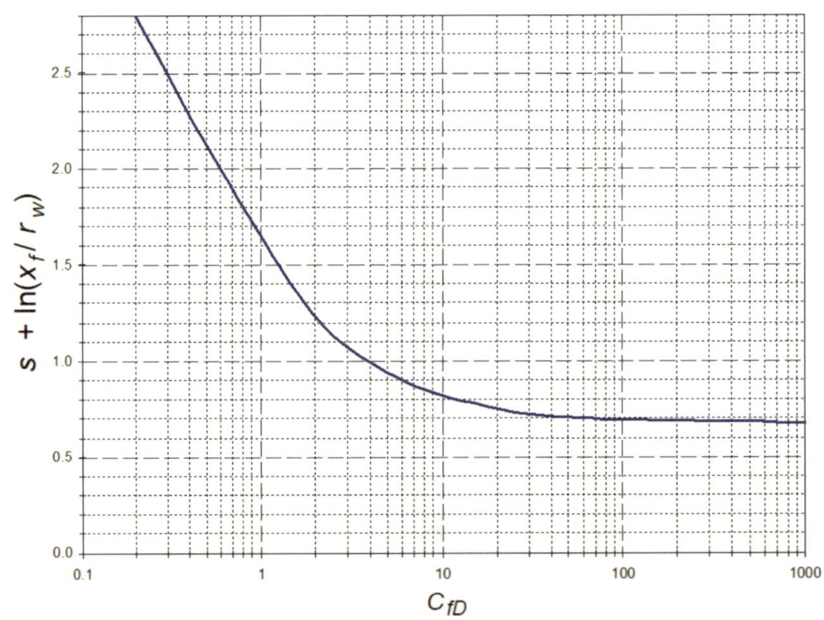

Figure 2.
Effect of skin factor on fracture conductivity (taken from [28]).

and

$$x_f = \frac{2.032qB}{h(t*\Delta P')_{L1}}\sqrt{\frac{\mu}{\xi\phi c_t k}}. \tag{24}$$

5. Points of intersection

If bilinear flow also takes place, then the point of intersection between the pressure derivatives of the bilinear and biradial flow lines, t_{BLBRi}, given by Eqs. (10) and (18), respectively, allows the development of an equation to find the half-fracture as follows:

$$\left(\frac{\pi t_{Dxf}}{26\xi}\right)^{0.36} = \frac{0.6125}{\sqrt{C_{fD}}}\left(\frac{t_{Dxf}}{\xi}\right)^{1/4}. \tag{25}$$

Simplifying,

$$\left(\frac{t_{Dxf}}{\xi}\right)^{0.11} = \frac{0.2862}{\sqrt{C_{fD}}}. \tag{26}$$

Replacing the dimensionless quantities, Eqs. (3) and (7) in Eq. (26) lead to

$$\left(\frac{0.000263kt}{\phi\mu c_t x_f^2\xi}\right)^{0.11} = 0.2862\sqrt{\frac{k\ x_f}{k_f w_f}}. \tag{27}$$

Solving for the half-fracture from Eq. (27), we readily obtain

$$k_f w_f = 10.5422\left(\frac{\xi\phi\mu c_t k^{3.5454} x_f^{6.5454}}{t_{BRBLi}}\right)^{0.22}. \tag{28}$$

By the same token, the intercept of Eq. (20) with Eq. (18), t_{LBRi}, provides another expression to find the half-fracture length:

$$x_f = \sqrt{\frac{kt_{LBRi}}{39.044\omega\phi\mu c_t}}. \tag{29}$$

Bilinear flow regime is absent in the plot of **Figure 1**. Linear, biradial, and radial flow regimes along with the late pseudosteady-state period are seen. The interception points formed by the possible combinations of such periods can be represented schematically in this plot.

Another way to find the half-fracture length comes from the intersection of Eqs. (2) and (10), t_{RBRi}, and Eq. (10) with Eq. (15), so that

$$x_f = \frac{1}{4584.16}\sqrt{\frac{kt_{RBRi}}{\xi\phi\mu c_t}} \tag{30}$$

and

$$x_f = 41.0554A^{1.3889}\left(\frac{\xi\phi\mu c_t}{kt_{BRPi}}\right)^{0.8889}. \tag{31}$$

The intercept point resulting between linear flow and bilinear flow lines given by the governing pressure derivative solutions, Eqs. (18) and (19), can be used to find either half-fracture length or permeability:

$$k = \left(\frac{k_f w_f}{x_f^2}\right)^2 \frac{16t'_{BLLi}}{13910\xi\phi\mu c_t}. \tag{32}$$

t_{BLRi} is the intersection of the bilinear pressure derivative line given by Eq. (18) with the radial flow regime line (Eq. (2)). This intersection point serves as the estimation of either permeability or fracture conductivity:

$$t_{BLRi} = 1677\frac{\xi\phi\mu c_t}{k^3}\left(k_f w_f\right)^2. \tag{33}$$

6. Other estimations

The expressions for determination of the naturally fractured reservoir parameters cannot be included in this chapter for space reasons. However, they can be found in [15, 16], which also used intersection points and maximum and minimum data read from the pressure and pressure derivative curve.

Radial flow regime may be absent in short tests run in fractured wells with the sole purpose of determining fractured parameters. For these cases, the skin factor can be estimated from any of the two empirical correlations presented by [27]

$$s = \ln\left[r_w\left(\frac{1.92173}{x_f}\right) - \frac{3.31739}{k_f w_f}\right] \tag{34}$$

and

$$s = \ln\frac{r_w}{x_f} + \frac{1.65 - 0.32u + 0.11u^2}{1 + 0.18u + 0.064u^2 + 0.005u^3}, \tag{35}$$

where

$$u = \ln C_{fD}. \tag{36}$$

Additionally, fracture conductivity can be read from the plot given in **Figure 2**.

Finally, space reasons prevent including *TDS* technique for fractured wells in unconventional shale formations. The reader is referred to [23–26].

7. Examples

7.1 Field example

[14] presented a field example of a fractured well test. Pressure and pressure derivative data are given in **Table 1** and **Figure 3**. Other relevant data are provided below:

$$q = 101\ \text{STB/D} \quad \phi = 0.08 \quad \mu = 0.45\ \text{cp}$$

$$c_t = 17.7 \times 10^{-6}\ \text{psia}^{-1} \quad B = 1.507\ \text{bbl/STB} \quad h = 42\ \text{ft}$$

$$r_w = 0.28\ \text{ft} \quad t_p = 2000\ \text{h} \quad P_i = 2200\ \text{psia}$$

$$\xi = 1$$

t, h	Δ*P*, psia	*t**Δ*P*', psia	*t*, h	Δ*P*, psia	*t**Δ*P*', psia
0.23	102	26.3	15	390	117
0.39	115	30	20	423	112
0.6	130	35.8	25	446	120
1	145	40.8	30	471	141
1.8	183	57.2	35	493	136.5
2.4	195	67	40	510	132
3.8	260	83.3	45	526	135
4.1	265	69.2	50	540	150
4.96	280	96.9	55	556	137.5
6.2	308	102.3	60	565	144
8.5	320	103.3	65	580	121.1
10	345	149	71	583	

Table 1.
Pressure data for field example (taken from [14]).

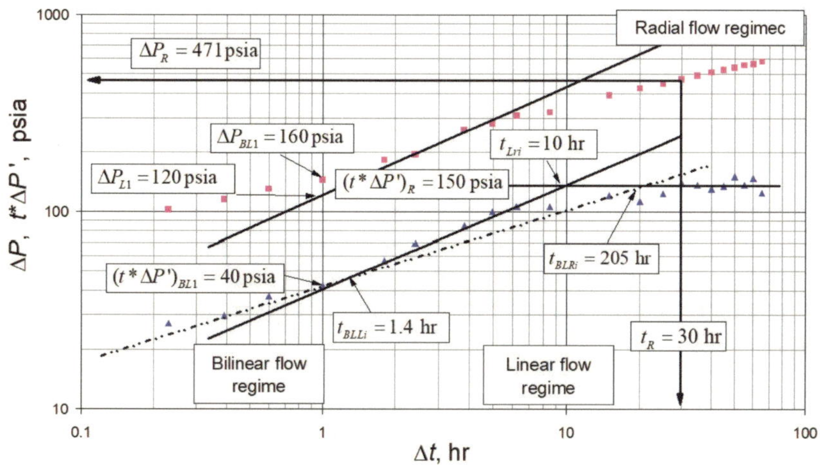

Figure 3.
Pressure and pressure derivative against time log–log plot for field example (taken from [14]).

Using a commercial well test software, the following parameters were estimated by nonlinear regression analysis:

$$k = 0.8 \text{ md}$$

$$xf = 82.2 \text{ ft}$$

$$k_f w_f = 300 \text{ md} - \text{cp}$$

The objective is to compute the hydraulic fracture parameters using the *TDS* technique and compare results obtained from the regression analysis.

7.1.1 Solution

7.1.1.1 Step 1: Obtain the characteristic points

Once the pressure and pressure derivative versus time log-log plot is built and reported in **Figure 3**, the characteristic points are read from such plot as follows:

$$t_R = 30\ \text{h} \quad \Delta P_R = 471\ \text{psia} \quad \left(t^*\Delta P'\right)_R = 150\ \text{psia}$$

$$\left(t^*\Delta P'\right)_{BL1} = 160\ \text{psia} \quad \Delta P_{BL1} = 40\ \text{psia} \quad \Delta P_{L1} = 120\ \text{psia}$$

$$t_{LRi} = 8.2\ \text{h} \quad t_{BLRi} = 195\ \text{h}$$

7.1.1.2 Step 2: Estimate permeability and skin factor

Permeability and skin factor are found in Eqs. (8) and (14) to be 0.76 md and −4.68, respectively.

7.1.1.3 Step 3: Estimate fracture conductivity

Fracture conductivity is estimated using Eqs. (21) and (22):

$$k_f w_f = \frac{121.74}{\sqrt{(0.08)(0.45)\left(17.7\times 10^{-6}\right)(0.76)}}\left(\frac{(101)(0.45)(1.507)}{(42)(40)}\right)^2 = 290.77\ \text{md-ft}$$

$$k_f w_f = \frac{1947.46}{\sqrt{(0.08)(0.45)\left(17.7\times 10^{-6}\right)(0.76)}}\left(\frac{(101)(0.45)(1.507)}{(42)(160)}\right)^2 = 290.7\ \text{md-ft.}$$

From **Figure 3**, t_{BLRi} = 200 hr. A very close value is obtained from Eq. (33):

$$t_{BLRi} = 1677\frac{(0.08)(0.45)\left(17.7\times 10^{-6}\right)}{(0.76)^3}(290.7)^2 = 205.71\ \text{hr},$$

which indicates that the calculation of the fracture conductivity is accurate. Notice that instead of estimating t_{BLRi} the fracture conductivity can be found instead to obtain another value of fracture conductivity; then, Eq. (33) can also be expressed as

$$k_f w_f = \sqrt{\frac{k^3 t_{BLRi}}{1677\xi\phi\mu c_t}} = \sqrt{\frac{0.76^3(205)}{1677(1)(0.08)(0.45)\left(17.7\times 10^{-6}\right)}} = 290.2\ \text{md-ft.}$$

7.1.1.4 Step 4: Half-fractured length and dimensionless fracture conductivity estimation

Find half-fracture length with Eqs. (23) and (24):

$$x_f = \frac{4.064(101)(1.507)}{(42)(120)}\sqrt{\frac{0.45}{(0.08)\left(17.7\times 10^{-6}\right)(0.76)}} = 79\ \text{ft},$$

$$x_f = \sqrt{\frac{k t_{Lri}}{1207\xi\phi\mu c_t}} = \sqrt{\frac{(0.76)(10)}{1207(0.08)(0.76)\left(17.7\times 10^{-6}\right)}} = 76.5\ \text{ft.}$$

Solve for half-fracture length from Eq. (13) and find this:

$$x_f = \frac{1.92173}{\frac{e^s}{r_w} - \frac{3.31739k}{w_f k_f}} = \frac{1.92173}{\frac{e^{-4.6844}}{0.28} - \frac{3.31739(0.76)}{290.7}} = 79\ \text{ft.}$$

Find the dimensionless fracture conductivity using Eq. (5):

$$C_{fD} = \frac{w_f k_f}{x_f k} = \frac{290.7}{79(0.76)} = 4.8.$$

The above value confirms that the fracture has finite conductivity.

7.2 Synthetic example

[13] presented a synthetic example of a pressure test run in a bounded homogeneous reservoir with the information given below:

$$B_o = 1.25 \text{ bbl/STB} \quad q = 300 \text{ STB/D}$$
$$h = 30 \text{ ft} \quad \mu = 5 \text{ cp}$$
$$r_w = 0.3 \text{ ft} \quad c_t = 3 \times 10^{-6} \text{ psi}^{-1}$$
$$P_i = 4000 \text{ psi} \quad \phi = 10\%$$
$$k = 33.334 \text{ md} \quad x_f = 200 \text{ ft}$$
$$A = 592 \text{ Acres}$$

Estimate the half-fracture length by the *TDS* technique, and compare the answer with the value used for generating the test.

7.2.1 Solution

7.2.1.1 Step 1: Obtain the characteristic points.

A pressure and pressure derivative versus time log–log plot is presented in **Figure 4**, from which the following characteristic points are read:

$$t_{BR} = 1.01 \text{ h} \quad \left(t^*\Delta P'\right)_{BR} = 64.63 \text{ psi} \quad t_{BRPi} = 3300 \text{ h}$$

7.2.1.2 Step 2: Half-fractured length estimation.

The half-fracture length is estimated with Eq. (12) and confirmed with Eq. (31), as follows:

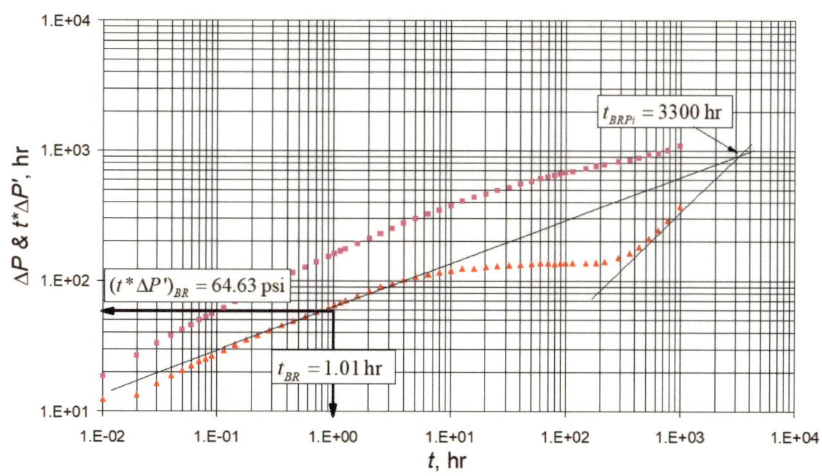

Figure 4.
Pressure and pressure derivative vs. time for synthetic example (taken from [13]).

$$x_f = 5.4595\left(\frac{300(1.25)}{30(64.63)}\right)^{1.3889}\sqrt{\frac{1.01}{(1)(0.1)(3\times10^{-6})}\left(\frac{5}{33.334}\right)^{1.778}} = 199 \text{ ft},$$

$$x_f = 41.0554(592\times43560)^{1.3889}\left(\frac{(1)(0.1)(5)(3\times10^{-6})}{33.334(3300)}\right)^{0.8889} = 201.6 \text{ ft}.$$

8. Comments on the results

The two given examples show three aspects of the *TDS* technique: (1) practical use, (2) accuracy, and (3) self-confirmation.

As shown in the exercises, the process includes defining flow regimes, drawing a few lines, and finally computing the necessary parameters. Contrary to the conventional straight-line method, which requires a plot for each flow regime, TDS technique uses only the pressure and pressure derivative versus time log-log plot. Computations are straight forward.

Table 2 summarizes the main parameters obtained in the two worked examples. The results show a good agreement between the calculated results by *TDS* technique and the results obtained from commercial software packages, for the field case. The results of the half-fracture length for the synthetic case using *TDS* technique are even better compared to the input value use to simulate the test. This demonstrates that *TDS* technique is an accurate methodology which has been also presented in many publications, not only in the list reference but also in others not mentioned here.

The last aspect dealt with is self-confirmation. In the field example, three values of half-fracture length and three values of fractured conductivity were found, and for the synthetic example, two values of half-fracture length were estimated from different equations. All the estimations match with the reference values.

	Field example						
		Obtained from					
Parameter	Commercial software	Eq. (21)	Eq. (22)	Eq. (33)	Eq. (23)	Eq. (24)	Eq. (13)
x_f, ft	82.2				79	76.5	79
$k_f w_f$, md-ft	300	290.77	290.7	290.2			
	Synthetic example						
		Obtained from					
Parameter	Commercial software	Eq. (12)	Eq. (31)				
x_f, ft	200	199	201.6				

Table 2.
Summary of results.

9. Conclusion

It has been shown that *TDS* technique is a powerful, practical, and accurate tool for well test interpretation because manipulations are easy to do and parameters can be confirmed from different sources from the same pressure test. Compared to reference values, the worked examples provided accurate results of both half-fracture length and hydraulic fracture conductivity. Besides being accurate, *TDS* technique has the great advantage of being able to estimate a given parameter, such as half-fracture length or fracture conductivity, from more than one source or equation. This provides a means of verifying that the estimated parameter is in a good range.

Acknowledgements

The author wishes to express his gratitude to Universidad Surcolombiana for providing him the time to write this chapter.

Nomenclature

A	Draining area (ft^2)
B	Oil volume factor (rb/STB)
C_{fD}	Dimensionless fracture conductivity
c_t	Compressibility (1/psi)
h	Formation thickness (ft)
k	Formation permeability (md)
$k_f w_f$	Fracture conductivity (md-ft)
P	Pressure (psi)
P_{wf}	Well-flowing pressure (psi)
q	Oil flow rate (STB/D)
q_g	Gas flow rate (MSCF/D)
r_w	Wellbore radius (ft)
x_f	Half-fracture length (ft)
s	Skin factor
t	Test time (h)
t_p	Production time (h)
$t^*\Delta P'$	Pressure derivative (psi)
$t_D{}^*P_D{}'$	Dimensionless pressure derivative

Greek symbols

Δ	Change
ϕ	Porosity (fraction)
λ	Interporosity flow parameter
μ	Viscosity (cp)
ξ	Variable to identify homogeneous ($\xi = 1$) or heterogeneous ($\xi = \omega$) reservoirs
ω	Dimensionless storativity coefficient

Suffixes

BL	Bilinear
*BL*1	Bilinear at 1 h
BLL	Bilinear-linear intersection
BR	Birradial
*BR*1	Birradial at 1 h
BRBLi	Birradial-bilinear intersection
BRPi	Birradial-pseudosteady intersection
D	Dimensionless
DA	Dimensionless based on area
Dxf	Dimensionless based on half-fractured length
DLBRi	Dual linear-birradial intersection
LBRi	Linear-birradial intersection

R	Radial
RBRi	Radial-birradial intersection
RPi	Intersect of radial-pseudosteady-state lines
w	Well
t	Time
P	Pseudosteady state

Author details

Freddy Humberto Escobar
Universidad Surcolombiana, Neiva, Huila, Colombia

*Address all correspondence to: fescobar@usco.edu.co

References

[1] Gringarten AC, Ramey HJ, Raghavan R. Applied pressure analysis for fractured wells. Journal of Petroleum Technology. Richardson, Texas, USA. 1975:793-800. DOI: 10.2118/5496-PA

[2] Cinco LH, Samaniego VF, Dominguez AN. Transient pressure behavior for a well with a finite-conductivity vertical fracture. SPE Journal. 1978;**18**(4):253-264. DOI: 10.2118/6014-PA

[3] Cinco-Ley H, Samaniego VF. Transient pressure analysis for fractured wells. Journal of Petroleum Technology. 1981; **Sept.**:1479-1766. DOI: 10.2118/7490-PA

[4] Cossio M. A Semi-Analytic Solution for Flow in Finite-Conductivity Vertical Fractures Using Fractal Theory. Richardson, Texas, USA: Society of Petroleum Engineers; 2012. DOI: 10.2118/163057-STU

[5] Tiab D. AAnalysis of pressure and pressure derivative without type-curve Matching: 1-skin and wellbore storage. Journal of Petroleum Science and Engineering. 1995;**12**:171-181. Also Tiab D. Analysis of pressure and pressure derivatives without type-curve matching: i-skin and wellbore storage. Society of Petroleum Engineers. 1993. DOI: 10.2118/25426-MS

[6] Escobar FH. Recent Advances in Practical Applied Well Test Analysis. New York: Nova Science Publishers, Inc.; 2015. 422p

[7] Tiab D. Analysis of pressure derivative without type-curve matching: Vertically fractured wells in closed systems. Journal of Petroleum Science and Engineering. Richardson, Texas, USA. 1994;**11**:323-333. DOI: 10.2118/26138-MS. This paper was originally presented as Society of Petroleum Engineers

[8] Escobar FH, Muñoz OF, Sepulveda JA. Horizontal permeability determination from the elliptical flow regime for horizontal wells. CT&F – Ciencia, Tecnología y Futuro. 2004; **2**(5):83-95. ISSN 0122-5383

[9] Escobar FH, Montealegre M. Conventional Analysis for the Determination of the Horizontal Permeability from the Elliptical Flow of Horizontal Wells. Richardson, Texas, USA: Society of Petroleum Engineers; 2007. DOI: 10.2118/105928-MS

[10] Escobar FH, Montealegre-M M. Determination of horizontal permeability from the elliptical flow of horizontal wells using conventional analysis. Journal of Petroleum Science and Engineering. 2008;**61**:15-20. ISSN: 0920-4105

[11] Escobar FH, Cantillo JH, Santos N. A practical approach for the estimation of the average reservoir pressure from multi-rate tests in long horizontal wells. Fuentes: El Reventón Energético. 2011; **9**(1):13-20

[12] Escobar FH, Montealegre MM, Cantillo JH. Conventional analysis for characterization of bi-radial (elliptical) flow in infinite-conductivity vertical fractured wells. CT&F – Ciencia, Tecnología y Futuro. 2006;**3**(2):141-147

[13] Escobar FH, Ghisays-Ruiz A, Bonilla LF. New model for elliptical flow regime in hydraulically-fractured vertical wells in homogeneous and naturally-fractured systems. Journal of Engineering and Applied Sciences. 2014;**9**(9):1629-1636

[14] Tiab D, Azzougen A, Escobar FH, Berumen S. Analysis of Pressure Derivative Data of Finite-Conductivity Fractures by the "Direct Synthesis" Technique. Society of Petroleum Engineers; 1999. DOI: 10.2118/52201-MS

[15] Escobar FH, Zhao YL, Fahes M. Characterization of the naturally fractured reservoir parameters in infinite-conductivity hydraulically-fractured vertical wells by transient pressure analysis. Journal of Engineering and Applied Science. 2015; **10**(12):5352-5362

[16] Tiab D, Bettam Y. Practical Interpretation of Pressure Tests of Hydraulically Fractured Wells in a Naturally Fractured Reservoir. Richardson, Texas, USA: Society of Petroleum Engineers; 2007. DOI: 10.2118/107013-MS

[17] Escobar FH, Tiab D, Berumen-Campos S. Well Pressure Behavior of a Finite-Conductivity Fracture Intersecting a Finite Sealing-Fault. Richardson, Texas, USA: Society of Petroleum Engineers; 2003. DOI: 10.2118/80547-MS

[18] Escobar FH, Zhao YL, Pournik M, Liu QG, Olaya-Marin G. Interpretation of pressure tests in uniform-flux fractured vertical wells with threshold pressure gradient in low permeability reservoirs. Journal of Engineering and Applied Science. 2015;**10**(20):9364-9372

[19] Escobar FH, Zhao YL, Zhang LH. Interpretation of pressure tests in horizontal wells in homogeneous and heterogeneous reservoirs with threshold pressure gradient. Journal of Engineering and Applied Sciences. 2014;**9**(11):2220-2228

[20] Escobar FH, Castro JR, Mosquera JS. Rate-transient analysis for hydraulically fractured vertical oil and gas wells. Journal of Engineering and Applied Science. 2014;**9**(5):739-749. ISSN: 1819-6608

[21] Al Rbeawi SJH, Tiab D. Effect of the Number and Length of Zonal Isolations on Pressure Behavior of Horizontal Wells. Society of Petroleum Engineers; 2011. DOI: 10.2118/142177-MS

[22] Escobar FH, Meneses AR, Losada LM. Straight-line conventional transient pressure analysis for horizontal wells with isolated zones. Dynamis. 2014; **81**(185):78-85

[23] Bernal KM, Escobar FH, Ghisays-Ruiz A. Pressure and pressure derivative analysis for hydraulically-fractured shale formations using the concept of induced permeability field. Journal of Engineering and Applied Science. 2014; **9**(10):1952-1958

[24] Escobar FH, Montenegro LM, Bernal KM. Transient-rate analysis for hydraulically-fractured gas shale wells using the concept of induced permeability field. Journal of Engineering and Applied Science. 2014; **9**(8):1244-1254

[25] Escobar FH, Rojas JD, Ghisays-Ruiz A. Transient-rate analysis for hydraulically-fractured horizontal wells in naturally-fractured shale gas reservoirs. Journal of Engineering and Applied Science. 2015;**10**(1):102-114

[26] Escobar FH, Bernal KM, Olaya-Marin G. Pressure and pressure derivative analysis for fractured horizontal wells in unconventional shale reservoirs using dual-porosity models in the stimulated reservoir volume. Journal of Engineering and Applied Sciences. 2014;**9**(12):2650-2669

[27] Tiab D. Advances in Pressure Transient Analysis — TDS Technique. Lecture Notes Manual. Norman, Oklahoma, USA: The University of Oklahoma; 2003. 577p

[28] Economides MJ, Watters, Dunn-Norman S. Petroleum Well Construction. New York: John Wiley & Sons; 1988. 622p

[29] Escobar FH, Gonzalez RA, Hernandez LM, Hernandez CM. Pressure and pressure derivative analysis for hydraulically fractured

vertical wells with face skin. Journal of Engineering and Applied Science. 2016; **11**(13):8268-8273

[30] Zhao YL, Escobar FH, Hernandez CM, Zhang CP. Performance analysis of a vertical well with a finite-conductivity fracture in gas composite reservoirs. Journal of Engineering and Applied Sciences. 2016;**11**(15):8992-9003

Chapter 5

Surface Drilling Data for Constrained Hydraulic Fracturing and Fast Reservoir Simulation of Unconventional Wells

Ahmed Ouenes, Mohit Paryani, Yamina Aimene, Chad Hammerquist and Aissa Bachir

Abstract

The objective is to present a new integrated workflow which leverages commonly available drilling data from multiple wells to build reservoir models to be used for designing and optimizing hydraulic fracture treatment and reservoir simulation. The use of surface drilling data provides valuable information along every wellbore. This information includes estimations of geomechanical logs, pore pressure, stresses, porosity and natural fractures. These rock properties may be used as a first approximation in a well-centric approach to geoengineer completions. Combining these logs from multiple wells into 3D reservoir models provides more value including using them in reservoir geomechanics, 3D planar hydraulic fracturing design and reservoir simulation. When using these 3D models and their results in a fast marching method simulator, the impact of the interference between wells can be estimated quickly while providing results like those derived with a classical reservoir simulator. Integrating surface drilling data with 3D reservoir models, hydraulic fracturing design and reservoir simulation into a single software platform results in a fast and constrained approach which allows for a more efficient management of unconventional wells.

Keywords: geomechanics, mechanical specific energy, laminated anisotropic rocks, interaction hydraulic and natural fractures, proppant transport, stimulated reservoir volume, fast marching method

1. Introduction

The recent findings of the Hydraulic Fracturing Test Site I industry consortium (HFTS 1) are well summarized in the September 2018 *Journal of Petroleum Technology* article titled "*Real Fractured Rock is So Complex it's Time for New Fracturing Models*" [1]. 600 ft. of core taken in a hydraulically fractured Wolfcamp reservoir in the Permian Basin, USA showed a more complex reality than what is accounted for in most hydraulic fracturing design and analysis software. This includes the interaction between hydraulic and natural fractures, which is largely ignored or poorly accounted for in most software currently used to model hydraulic fractures and their resulting geometry. Rassenfos [1] emphasized in his summary of multiple recent publications describing HFTS 1 findings, that the "*fracture height is overrated.*

While microseismic testing indicated that fractures grew up about a 1000ft, the height of the propped fractures- the fractures most likely to produce oil and gas was about 30ft." Rassenfos [1] summary article discusses the important role played by the natural fractures but misses multiple other challenges facing the realistic modeling of hydraulic fracturing. Among the most noticeably and urgent challenges is the lack of data to better characterize the key inputs needed by any hydraulic fracturing modeling approach and the role of interfaces and their impact on vertical fracture growth.

The nature of the unconventional revolution, and ensuing extensive use of hydraulic fracturing, is the prevalent belief that the majority of the wells need to be drilled and stimulated in a "factory mode" where useful data such as wireline logs are not acquired at a statically significant rate. This philosophy, exhibited by many major unconventional players, has left a void of data in most major fields, greatly undermining optimization efforts necessary to economically produce said fields. A solution exists in the use of surface drilling data, which is available at every well. These drilling data are acquired by all drilling contractors around the world and are used qualitatively and quantitively during drilling operations. The surface drilling data include torque (T), rate of penetration (ROP), weight-on-bit (WOB). Since this drilling data is available at any old, current and future well and most operators dealing with unconventional reservoirs are not acquiring wireline logs at all the wells, the authors investigated the possibility to estimate pseudo-logs from surface drilling data. Using surface drilling data to infer rock properties, pore pressure, and stresses, comes with multiple challenges, which when overcome, open the door to improvements to the physics used in modeling hydraulic fracturing.

2. Newly discovered value in surface drilling data and mechanical specific energy (MSE)

Ouenes et al. [2] introduced the use of surface drilling data to simultaneously estimate the rock geomechanical properties, pore pressure, stresses, porosity and natural fractures needed to guide the steering of horizontal wells within the most frackable rock in real time, and additionally provide a completion design for optimal hydraulic fracturing when drilling is finished. The Mechanical Specific Energy (MSE) [3] computed from commonly available surface drilling data such as torque (T), rate of penetration (ROP), weight-on-bit (WOB) and bit diameter (D) has been widely used to improve drilling efficiency. The Mechanical Specific Energy is defined as

$$\mathrm{MSE} = 4\left(\frac{\mathrm{WOB}}{\pi \mathrm{D}^2}\right) + \left(\frac{480\mathrm{N} \times \mathrm{T}}{\mathrm{ROP} \times \mathrm{D}^2}\right) \tag{1}$$

All the components used in the MSE equation are commonly measured at the surface during drilling operations. Most horizontal shale wells currently being drilled use a drilling motor which requires the use of a different term for the Rotation Speed N^*. When using a motor, the MSE requires the use of the formula given below in Eq. (2) where N is the rotational speed of the drill pipe, Kn is the mud motor speed to flow ratio and Q is the total mud flow rate.

$$N^* = N + K_n \times Q \tag{2}$$

Once the MSE is available, it can be used for multiple purposes including deriving geomechanical properties, pore pressure, stresses, porosity, and natural fracture indicators. To fully take advantage of the derived MSE, it should be further combined with other drilling information. Once MSE is available, the Unconfined

Compressive Strength (UCS) can be derived from it in multiple ways. However, the UCS derived from the MSE needs to be corrected as the bit performance is also significantly influenced by the differential pressure.

2.1 Corrected MSE

Most of the recent MSE applications for completion optimization use surface drilling data which do not represent the MSE at the drill bit. The challenge posed by the use of surface drilling data consists of finding a way to eliminate costly and risky downhole equipment to measure the downhole MSE while ensuring accurate results. The solution for this challenge is to correct the surface drilling data by removing the frictional losses along the borehole. The Corrected Mechanical Specific Energy (CMSE), which is calculated in real time and uses surface drilling data, wellbore geometry, and drilling equipment parameters to estimate the friction losses along the drill string, was shown [4–6] to be a viable solution. This new technology uses advanced drilling and wellbore mechanics to estimate the multiple factors that create the frictional losses in real time and has been validated in multiple wells and basins.

Commonly available surface drilling data such as torque, weight on bit, rate of penetration and RPM are used as inputs to the model, along with the mud motor differential pressure and flow rates. Mud motor specifications, such as the maximum limits of differential pressure and flow rates, are also important inputs to normalize the computed Mechanical Specific Energy. When a Rotary Steerable Tool (RST) is used to steer the well, the Bottom Hole Assembly (BHA) information is a necessary input along with the wellbore survey to accurately perform the torque and drag analysis and to estimate the friction pressure losses along the wellbore. The friction pressure losses are then extracted from the surface MSE to compute the Corrected MSE.

The Drilling Efficiency (DE), which is the ratio of the energy required over energy spent in breaking a unit volume of the rock, is computed based on the CMSE and the Confined Compressive Strength (CCS) as shown below in Eq. (3)

$$DE = \frac{\text{CCS}}{\text{CMSE}} \tag{3}$$

As shown in Eq. (4) CCS accounts for the typically increasing Unconfined Compressive Strength (UCS) rock strength with depth as well as the effects of the confining stresses (Δp) and angle of internal friction factor (θ) applied on the rock. By correlating the MSE with CCS through the DE and by fitting a trendline on the computed DE data-set, the pore pressure can be estimated by accounting for the variations of DE data-set from the DE trendline. The fit of the DE trendline should be calibrated with pore pressure measurements from DFIT tests and the DE trendline should be updated accordingly.

$$CCS = \text{UCS} + \Delta\text{p}\left(\frac{1 + \sin\theta}{1 - \sin\theta}\right) \tag{4}$$

Once these friction losses are correctly estimated, they can be used to correct the MSE measured from surface drilling data which can be compared to measured downhole MSE. The principle of the predictive model is that torque and drag forces in a directional wellbore are primarily caused by sliding friction. Sliding friction force is calculated by multiplying the sidewall contact force with a friction coefficient. A lumped-parameter model provides the basis for the prediction of torque and drag. Both torque and drag are caused entirely by sliding friction forces that result from contact of the drill string with the wellbore. The frictional forces are subtracted from the surface MSE to accurately estimate the corrected MSE.

2.2 Comparing corrected MSE from surface drilling data to downhole measured MSE

Figure 1 shows an example from the Gulf of Mexico data set where Majidi et al. [7] emphasized the importance of using the downhole torque measurements as the torque term in the MSE often dominates the WOB term (see Eq. (1)). The discrepancy between MSE from uncorrected surface data and from the use of downhole measurements is clear in **Figure 1**. Using all the available drilling information, the frictional losses are estimated in order to compute the CMSE. The similarity between Majidi et al. [7] MSE from downhole measurements and the results computed from corrected surface drilling data is illustrated in **Figure 1**. Using the CMSE approach where surface drilling data are corrected for friction losses, results qualitatively comparable to measured downhole MSE can be derived. This opportunity leads to a reasonable estimation of mechanical rock properties at any well using commonly available surface drilling data thus circumventing the major problem of lack of well data in unconventional reservoirs.

2.3 Estimating rock properties from CMSE

The next step is to leverage the estimation of CMSE and UCS to build a real-time wellbore geomechanical model. The CMSE is directly used as a proxy for UCS by finding a linear correlation of the CMSE to the average UCS values in the zone of interest. Velocity in rocks primarily depends on three factors namely porosity/effective pressure, saturating fluids and lithology/rock minerals. When focusing primarily on the lateral section, it is reasonable to assume that saturating fluids are fairly homogenous. Thus, the two contributing factors to acoustic and shear velocity become lithology and porosity/effective pressure which are used to estimate these velocities and the rock mechanical properties. For example, the Young's Modulus (YM) can be derived from UCS using multiple available correlations based on different lithologies. Knowing the YM could lead to using other correlations to estimate the Poisson's ratio (PR), Shear Modulus (G), Porosity (PHI), Fracture Index (FI), and rock brittleness (STRBRT). Majidi et al. [7] showed how the MSE could also be used to derive pore pressure. Using frictional faulting theory, with

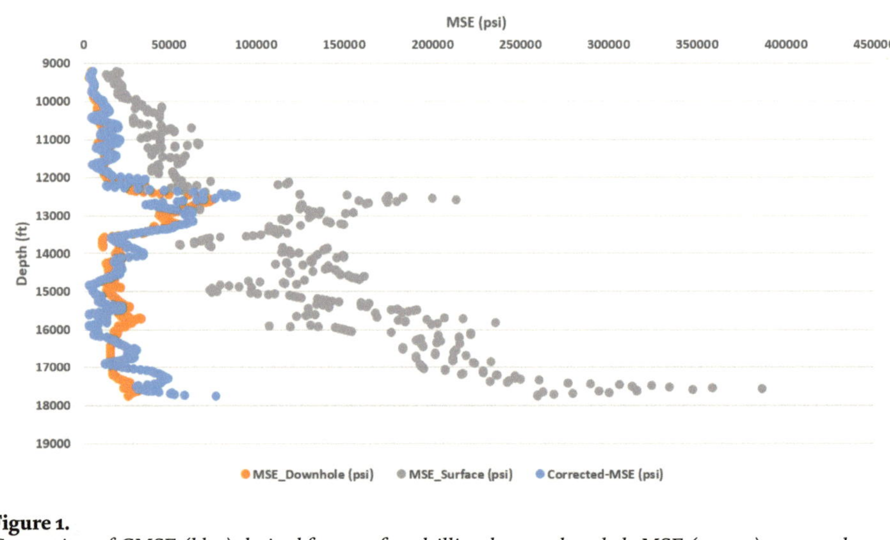

Figure 1.
Comparison of CMSE (blue) derived from surface drilling data vs. downhole MSE (orange) measured downhole. Notice the difference of the CMSE values from the MSE derived from surface drilling data without correction (gray).

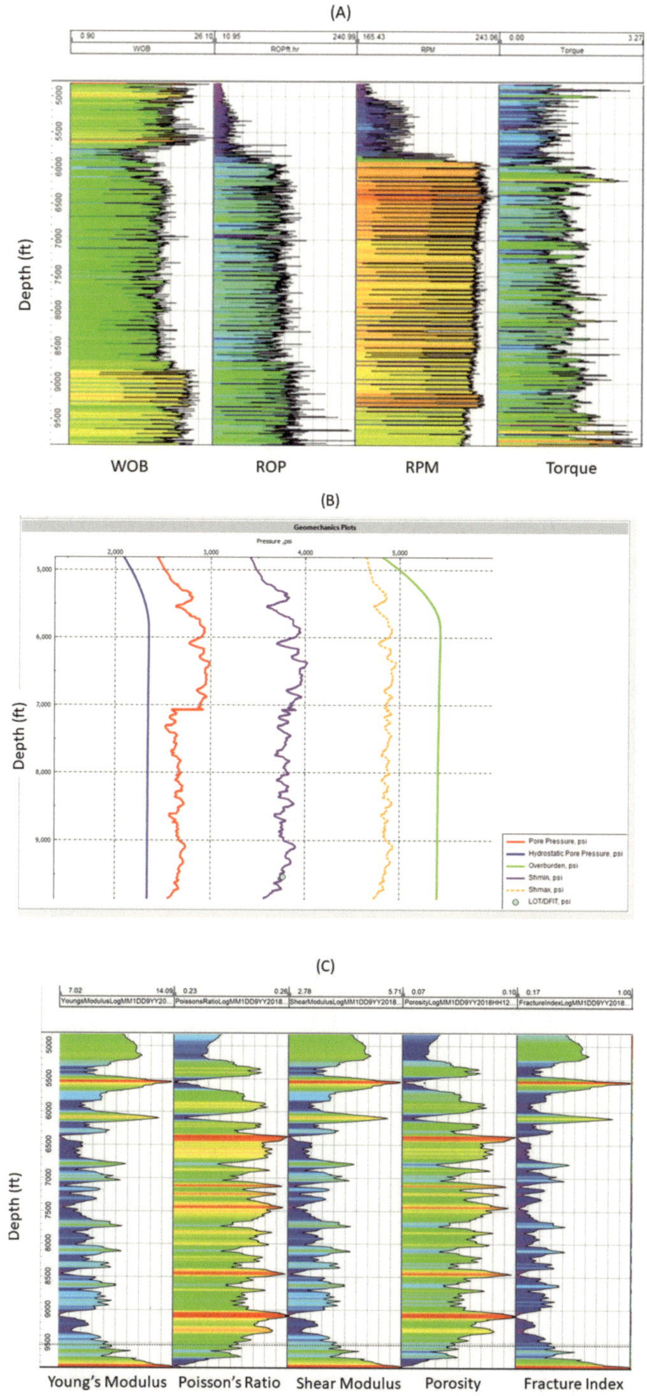

Figure 2.
(A) Using the commonly found surface drilling data to estimate the (B) pore pressure and stresses and (C) key geomechanical logs, porosity and natural fractures along any wellbore.

the UCS and the pore pressure, in-situ stresses can also be estimated. **Figure 2(A)** illustrates the common input surface drilling data and the resulting outputs that include stresses in **Figure 2(B)** and rock properties in **Figure 2(C)**. Using these key rock properties as inputs, multiple other properties combining both rock properties and stresses can be derived and used in completion optimization.

3. Engineered completion using surface drilling-derived logs in every unconventional well

Production diagnostic tools have time and time again confirmed the variable performance of stimulated stages, prompting the need to geoengineer completions to adapt the treatment to the variable nature of the stress and rock properties along the wellbore. Fortunately, the use of surface drilling data leads to the estimation of key rock property logs, minimum stress and CMSE that could be used as a reference log to geoengineer the different hydraulic fracturing stages. The objective is to set the stages and the clusters in rock with similar fracture gradients, so they break at a common treating pressure. **Figure 3** shows a well that has benefited from the use of the CMSE derived from surface drilling data and the variable reference log used to geoengineer the completion. When starting with a geometric design an initial cluster efficiency of 62% is computed along the wellbore. As shown in **Figure 3** in the two red rectangles, minor changes in the position of the stages and their clusters could increase the cluster efficiency to 70%. Current applications of this technology to various wells have shown a consistent increase from 10 to 30% of cluster efficiency when using the logs derived from surface drilling data. Fiber optic measurements have confirmed the ability of these surface drilling derived logs to predict the cluster efficiency within a reasonable engineering error. While fiber optic measurements are extremely rare and very expensive, on the order of a million dollars, the logs derived from surface drilling are available at every well at an insignificant cost.

Given the density of unconventional wells, this newly available information provides the opportunity to go beyond the wellbore and start propagating this well-centric information into a 3D representation of the reservoir.

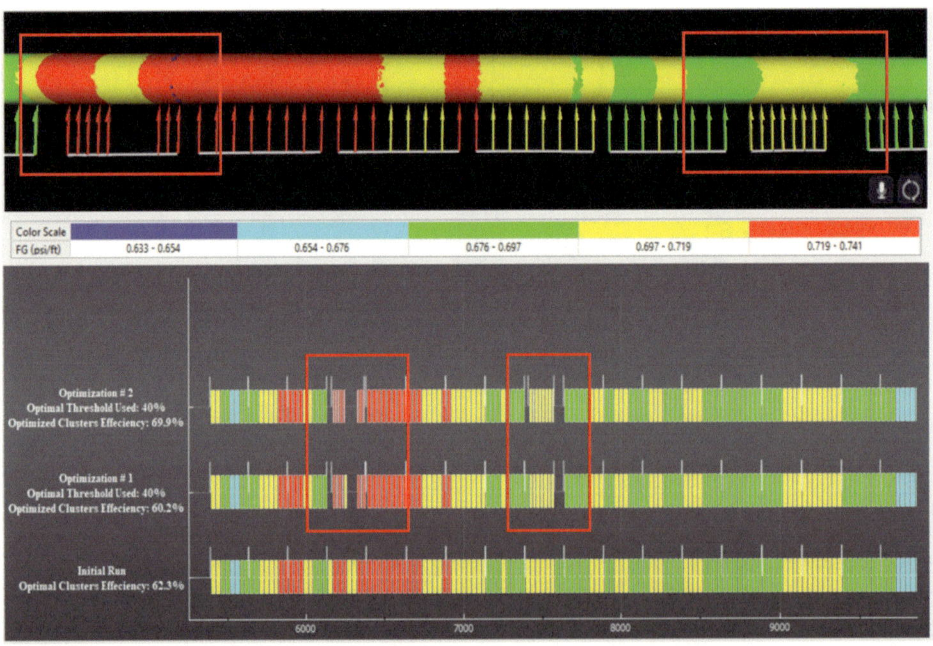

Figure 3.
Using a surface drilling derived reference log to geoengineer completions by moving stages to target similar rock, increasing cluster efficiency.

4. Well-based 3D modeling using geostatistics and artificial intelligence: propagating the surface drilling derived logs across the reservoir volume

In Sections 2 and 3, it is shown how the rock mechanical properties, pore pressure and stresses are extracted along the lateral section of the wellbore and are used as reference properties to engineer the completion to improve the overall efficiency. In this section, these properties are propagated in 3D space to accurately characterize the reservoir, so that these inputs can be fed into the hydraulic fracturing design and reservoir simulation workflows explained in the subsequent sections. The large number of wells drilled in unconventional assets combined with the estimation of critical logs at all the wells from surface drilling data provides the opportunity to propagate the well information into a 3D reservoir model. Since many companies do not have seismic on their acreage or for budgetary reasons do not plan to license the existing seismic, these multiple logs derived at all of the wells allow the construction of reliable 3D reservoir models. These 3D models could be estimated in a stratigraphic framework over a large area that encompasses many wells. In such cases, geostatistics could be used to estimate the distribution of gamma ray, porosity, Young's Modulus, Poisson's ratio and shear modulus. However, the pore pressure, minimum stresses and natural fracture are more complex continuous properties that need to be estimated with neural networks [8, 9] and other artificial intelligence tools able to capture the complex geology that control their variability.

One major reason for propagating these rock properties in 3D is to provide that information to a 3D planar hydraulic fracturing simulator as well as to geomechanical software. To achieve this goal, all the wells are used together in a large reservoir grid to create the 3D models from which smaller well grids (**Figure 4A**) will be extracted around a well or a pad. With this approach, all the available well data will be used to improve the 3D distribution of the key properties needed for the 3D

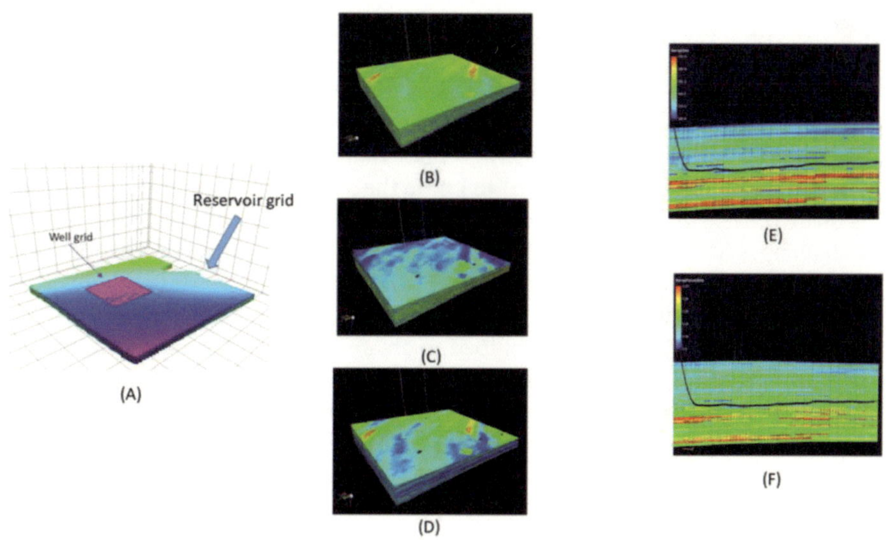

Figure 4.
(A) A large stratigraphic 3D geocellular grid is built from all the available wells to propagate 3D reservoir models. A smaller, higher resolution grid is extracted around the well that provides to the 3D planar hydraulic fracturing simulator: (B) Young's Modulus, (C) Poisson's ratio, and (D) unconfined compressive strength (UCS). (E) A cross section in the well grid of the minimum stress and (F) Poisson's ratio honoring the lateral and vertical variability captured by the 3D models."

planar hydraulic fracturing simulator. The other benefit of these derived 3D models is the estimation of the stress gradients resulting from the interaction between regional stresses and the three sources of stress perturbation created by the local geology: (1) variable geomechanical properties, (2) pore pressure and (3) natural fractures all available from the extrapolation in 3D of the logs derived from surface drilling data. Because of their importance in estimating these stress gradients, the modeling of the natural fractures requires some particular attention.

5. The importance of 3D natural fracture models

Natural fractures have a significant impact in unconventional reservoirs, yet they are rarely accounted for in most physical modeling related to hydraulic fracturing. Natural fractures could have a positive impact as they create additional surface contact during hydraulic fracturing which is commonly referred as fracture complexity [10]. This can be predicted with geomechanical modeling and validated with microseismic response [11, 12]. The contribution of the natural fractures could also be negative by creating direct links to water bearing faults [13] or by creating frac hits [14, 15], through poroelastic effects, that will often damage the production from child and parent wells. Given their importance, a predictive model that provides the 3D distribution of these natural fractures is a critical input for any model trying to predict the outcome of hydraulic fracturing. However, finding a 3D distribution of the natural fractures has two major challenges: how to define the natural fractures at the wells given the rare occurrence of core or image logs in unconventional wells, and how to distribute the limited well data in the 3D reservoir to create a predictive model.

One of the motivations behind the use of surface drilling data is to be able to extract a fracture indicator that can be used to enrich the poor and limited statistics of natural fractures indicators found at wells and using these to build a 3D natural fracture model. Since the natural fractures are not a result of a depositional phenomenon only, their prediction is very different from mapping a more conventional property such as reservoir porosity. Having few limited wells with core or image logs will likely not provide the full statistics of these natural fractures. In other words, most statistical methods such as Discrete Fracture Networks (DFN) may not have the proper statistics from the wells to make any reasonable prediction of their 3D distribution. Attempts were made to reduce this problem by constraining the DFN with a continuous property [16] to guide the statistical distribution but other issues made the use of DFN in natural fracture modeling very challenging. Among these challenges include the dramatic variations found in the upscaled properties [17] needed for additional use of the DFN in engineering applications.

To avoid all the issues related to the use of a DFN, the Continuous Fracture Modeling (CFM) approach was developed to create validated predictive models of natural fractures [8, 9]. The CFM approach takes full advantage of the surface drilling derived fracture index or even the limited statistics that can be found in any natural fracture proxy, image log or core. The CFM approach honors structural geology concepts and focuses on the drivers that influence the presence of natural fractures. For example, the density of natural fractures at a given point in the reservoir does not depend on poorly sampled statistics of various fracture sets measured through limited wireline data, but on the volumetric distribution and interaction of lithology, structural settings and distance to faults, porosity, and many other reservoir properties that create the resulting natural fractures. These reservoir properties commonly called natural fracture drivers could all be estimated directly or indirectly through geologic modeling and seismic processes that involve seismic inversion, spectral decomposition and volumetric curvatures [9] when the data is available. Since the relationship between the natural fracture drivers and the limited natural fractures available at the wells is complex, artificial intelligence [8] is used not only to retrieve any existing and potential correlations that honors the limited statistics but also all the structural geology concepts. With this approach, extrapolation beyond the limited statistics is possible and has been successfully applied during the last three decades to various problems requiring an accurate description on where the natural fractures are. Among these, are problems in geomechanics such as interactions between hydraulic and natural fractures. This interaction could be better understood if studied in a decoupled way; where the natural fractures role in altering the regional tectonic stress and their impact on the lateral propagation of the hydraulic fracture is separated from the effects of natural fractures during the vertical propagation of the hydraulic fractures.

6. Constraining the hydraulic fracture propagation in the horizontal direction

A major shortcoming in most of the current hydraulic fracture simulators is the assumption that shale reservoirs at the scale of the wellbore are subjected to a homogeneous stress environment. Hence, hydraulic fracture stages were designed based on a constant stress field, in terms of magnitude and orientation. Unfortunately, lateral stress gradients, and their effects on microseismicity, have evidenced a more complex situation. The origins of these lateral stress gradients are numerous and include variation of the rock geomechanical properties, pressure depletion around existing wells, and proximity to faults and their associated natural fracture systems.

A decoupled approach using a plane strain framework to capture the lateral stress gradients was used by Aimene and Ouenes [11]. Their geomechanical modeling uses as input the three key factors affecting the lateral stress gradients: rock elastic properties, reservoir pressure, and natural fractures. The elastic properties and reservoir pressure models derived in previous sections from surface drilling data are used as inputs for the geomechanical model. The model uses explicit fractures to describe the distribution of the natural fractures then simulates the proper initial stress conditions resulting from the various sources of stress variability followed by the simulation of hydraulic fracturing in this heterogeneous stress medium. Since microseismic data is limited to only a few wells, the geomechanical approach used to capture the lateral stress gradients must be able to predict microseismicity rather than use it as calibration. The resulting geomechanical simulation predicts the differential stress, stress rotations and strain which serves as a reasonable proxy for microseismic events for validation as shown in **Figure 5**.

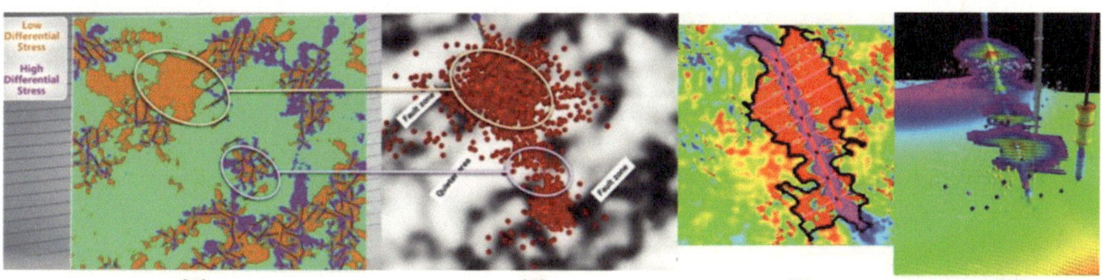

Figure 5.
Differential stress (A) and strain (C) validated with microseismicity (B) and the resulting geomechanically constrained hydraulic fractures (D).

This geomechanical approach combines the advantages of the particle-based numerical method, Material Point Method (MPM), a meshless numerical method, and the CFM approach to solve for a general continuum mechanics problem where all reservoir realities (natural fractures, variable rock properties and reservoir pressure heterogeneity), can be accounted for when estimating the stress field prior to stimulation and its subsequent perturbation during hydraulic fracturing. A direct benefit of this geomechanical approach is the quick estimation of the differential stress.

6.1 Estimation of differential stress: geomechanics vs. surface drilling data

The first result of the reservoir geomechanics approach [11] is the differential stress (**Figure 6**) which can be used as shown in Paryani et al. [18] to geoengineer completions. The advantage of using differential stress for geoengineering completions is the ability to consider the complex geology beyond the wellbore. In other words, well-centric approaches such as the one relying entirely on using a reference log derived from surface drilling data, are approximations that work only if the geology is not highly variable around the considered well. When the geology is variable with significant variability of the geomechanical properties, natural fractures, and pore pressure, then the best approach is to use the derived 3D models as input in the reservoir geomechanics approach [11] to estimate the differential stress.

Figure 6 shows a two well pad where faults (**Figure 6A**) were interpreted from multiple wells and used as input in the geomechanical approach [11] to compute the differential stress (**Figure 6B**). Since the analysis of surface drilling data was available in both wells, the differential stress was also estimated by using only that limited information. The computed differential stress from surface drilling data (**Figure** 7, right track) is compared to the one derived from the reservoir geomechanical simulation (**Figure** 7, left track) extracted along the wellbore from the differential stress distribution shown in **Figure 6B**. This comparison shows very strong similarities between the differential stress derived from full reservoir geomechanics (**Figure** 7, left track) with the one derived from the well centric approach based

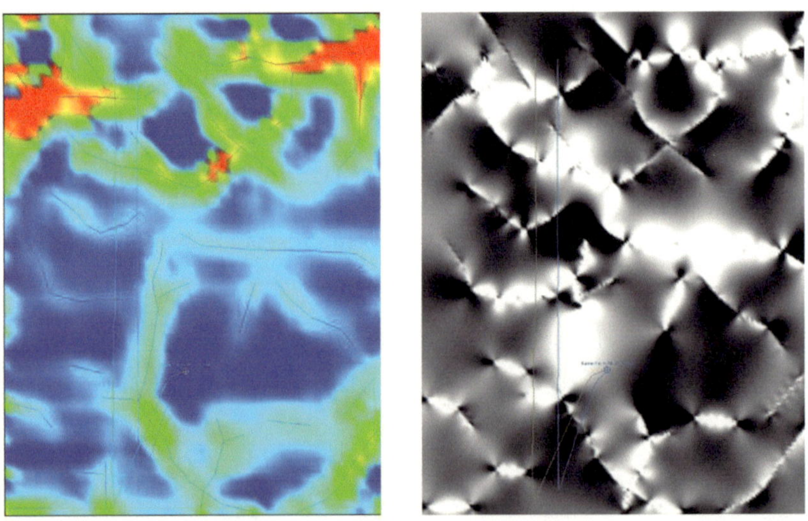
(A) (B)

Figure 6.
(A) Interpreted faults used as input in the reservoir geomechanics that estimates (B) the differential stress and the lateral stress gradients needed to geoengineer the stages.

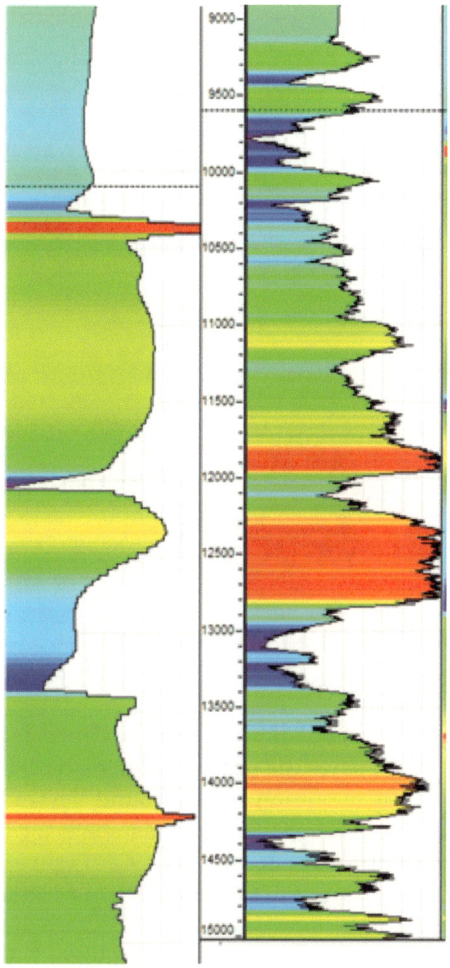

Figure 7.
(Left) Differential stress derived from reservoir geomechanics results shown in **Figure 6B**. *(Right) differential stress derived from a well centric approach using only surface drilling data.*

only on surface drilling data (**Figure 7**, right track). Both curves indicate the same zones for high differential stress zones where engineered completions are required to overcome earth resistance to hydraulic fracturing. The engineered completion could adjust the pumping parameters, stage length and number of clusters according to the derived differential stress with the objective of pumping bigger stage lengths in areas of low differential stress and vice versa. Areas of low differential stress will promote complex fracturing whereas areas of high differential stresses will result in planar hydraulic fractures with lower cluster efficiency as shown in **Figure 8**. The resulting engineered completion can be derived within a few hours of the well reaching Total Depth (TD) which illustrates the benefits of using surface drilling data if no other information is available.

6.2 Estimation of stress rotation

The knowledge of differential stress is important to predict the ability of creating fracture complexity and increased surface contact. Areas with higher differential stress will produce highly anisotropic hydraulic fractures with reduced surface contact. This problem will be further complicated if the maximum horizontal stress

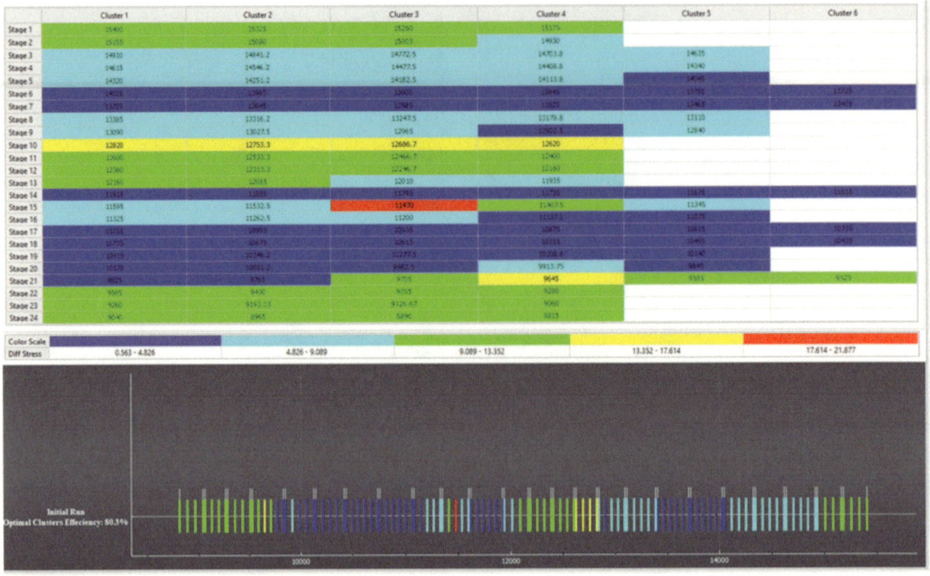

Figure 8.
Engineered completion using the differential stress as a reference log to adjust stage length and number of clusters.

S_{Hmax} direction is not locally perpendicular to the wellbore, leading to undesirable parallel hydraulic fractures. Hence the need to estimate both the differential stress and the local direction of the maximum stress around the wellbore.

A good example that can illustrate this critical issue is the Grisham fault in the Permian basin, USA where stress rotates by up to 90°. **Figure 9** shows the public domain faults [19–20] used as input in the geomechanical simulation [11] to estimate the stress orientation.

When the Woodford faults are subjected to a dominant E-W tectonic stress, rotations in the maximum horizontal stress direction arise as shown in **Figure 9**. It is important to note that faults which are oriented parallel or perpendicular to in-situ maximum stress direction, such as the N-S trending fault, cause little perturbation and critically stressed faults (roughly 30–60° from local maximum stress direction) cause large perturbations. While much of the basin is still subject to a S_{Hmax} within 10° of the input orientation, several areas evidence large deflections from this input orientation. These deflections can be further refined if using seismic data to define the local faults. Given the variability of the differential stress and the direction of the maximum stress direction, both captured with the plane strain modeling the next step is to evaluate the actual strain resulting from the hydraulic fracturing.

6.3 Estimation of strain for laterally constrained 3D planar hydraulic fractures

The 2D plane strain MPM modeling provides valuable stimulated reservoir volume (SRV) information by modeling the effect of a large increase in the stress around the wellbore and its distribution throughout the reservoir volume and interaction with fractures and faults as well as accounting for any variable geomechanical properties and pore pressure of the rock. This is achieved numerically by applying a large pressure on a hydraulic fracture plane with a given length varying between 100 and 200 ft. which is used to model the effects of the pumping pressure in the reservoir. The real surface contact available to the fluid to apply its pressure and create a stress front is much larger than the numerical hydraulic fracture assumed to be around 150 ft. Thus, the pressure applied to this limited surface must be higher than

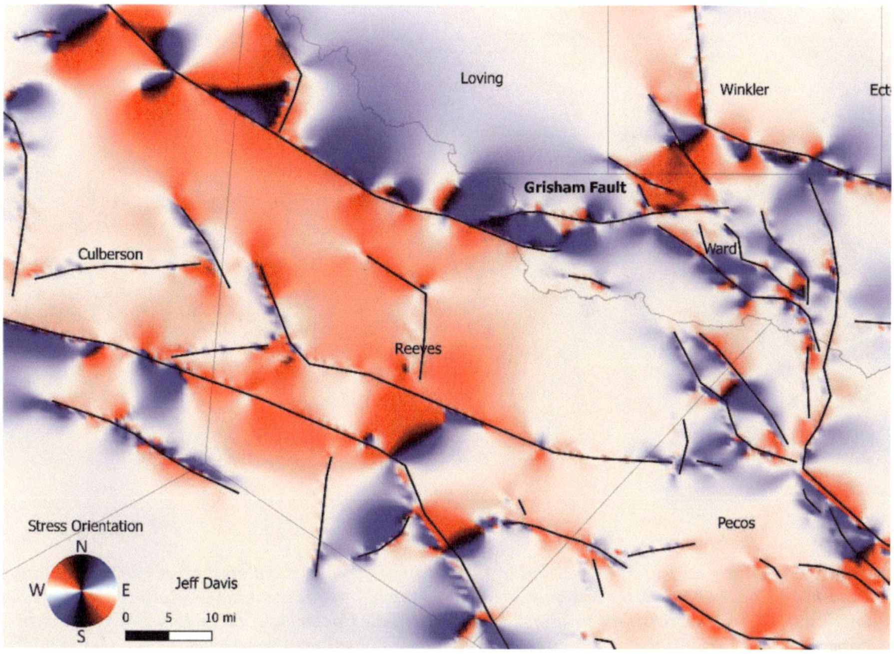

Figure 9.
Stress orientation around the Grisham fault showing distinct behaviors of the stress field orientation north and south of the fault, and also along strike of the fault.

the pumping pressure and is approximately, in most realistic unconventional wells, about 2.5 times the minimum stress value. Since this stress can be modeled with a dynamic simulation, the pressure applied to the hydraulic fractures can be applied sequentially, in parallel, or in a zipper mode. This ability to simulate the sequence of hydraulic fracturing allows the proper representation of stress shadow effects between stages as well as those seen between wells. These stress shadowing effects are considered along with the complex geology present between the stages and wells. For each hydraulic fracturing sequence, the resulting strain will be able to provide useful indication on the resulting SRV as shown in **Figure 10**.

One simple way to account for the lateral stress gradients captured by the geomechanical simulation, is to estimate the geomechanical half lengths (**Figure 5C**) from an interpreted envelope of the strain (**Figure 10B**) that could represent a proxy for the SRV. These interpreted asymmetric geomechanical half lengths are used at each cluster or stage as a constraint in a 3D planar hydraulic fracture design. It is important to reemphasize that the use of the planar representation of the hydraulic fractures is not an indication that the hydraulic fractures are indeed planar but a simple mathematical discretization of an SRV estimated by the full geomechanical simulation.

Having a constraint in the lateral direction is very helpful for a better estimation of the fracture height when using a 3D planar hydraulic fracturing approach. In this model, the vertical fracture growth occurs in the simplified world of perfect interfaces where debonding does not occur in a layered anisotropic rock. Unfortunately, the fracture growth does not depend only on the lateral stress but also on the geologic nature of the laminations and the characteristics of their interfaces which could be weak and could shear and consume hydraulic fracturing energy thus reducing the hydraulic fracture height. Since we have successfully estimated and validated with microseismic data the lateral stress gradients estimated with 2D plane strain MPM model, this information can be used as an input in a 2D vertical problem where we will focus on the geologic factors affecting the vertical fracture growth.

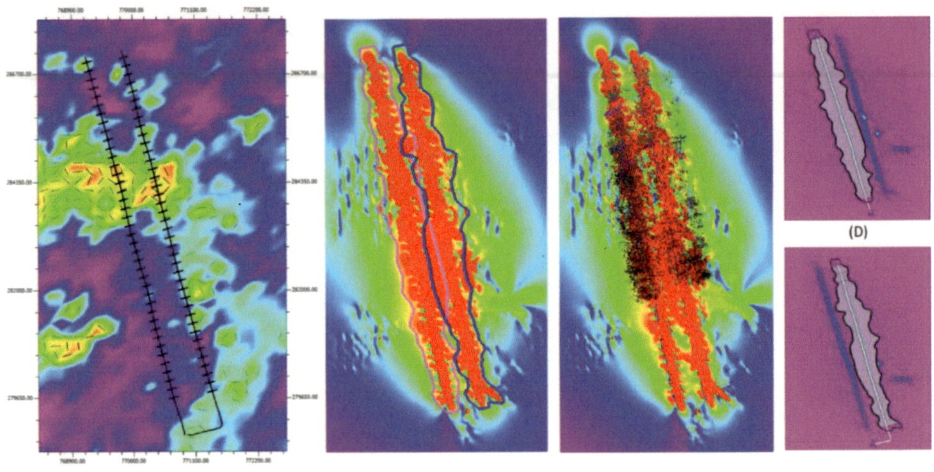

Figure 10.
(A) Equivalent fracture model (EFM) derived by CFM using only surface drilling derived fracture indicators logs. The EFM is used as input in the reservoir geomechanics that provides the initial perturbed stress field and the subsequent (B) asymmetric strain resulting from the hydraulic fracturing of the wells and (C) comparison to microseismic events (note: the heel section of the wells was not monitored due to operations). (D) and (E) The envelope of strain provides the gross geomechanical half lengths which provide the lateral stress gradients needed to constrain the 3D planar hydraulic fracturing simulator.

7. Constraining the hydraulic fracture propagation in the vertical direction in the presence of weak interfaces and natural fractures

Interfaces are among the geological features that are known to have an impact on the vertical propagation pattern and the final fracture height in unconventional reservoirs. Interfaces limit adjacent lithologies with similar or contrasting properties. It could be a material or just contact between two adjacent lithologies. Typical material thickness is between 1 and 500 mm for volcanic ash layers, and μm to mm thickness for mineralized veins, highly or partially mineral filled fractures, organic matters layers in the form of bitumen lubricating film or kerogen coating the surface of the interface [21] and bentonite layers [22] than could vary in thickness up to a few centimeters.

Interface mechanical properties can either be strong or weak and can be further weakened by tectonic deformations. They are a source of displacement discontinuities and delamination and are associated with fracture propagation behaviors like kinking, offsets, bifurcation, stepping over and termination. This suggests that displacement continuity hypothesis that are used in many hydraulic fracturing models where contact mechanics between layers with stick conditions (no sliding) is used to model the interface between layers [23, 24], is not valid for modeling real interface effects. In fact, using stick condition within a stratified structure could not account for sliding and de-cohesion between the layers due to hydraulic fracture pressurization. Thus, the need to use a proper interface model that (i) accounts for the displacement discontinuity between layers and (ii) allows the decohesion and interface delamination at the interface.

Explicitly modeling the interfaces presents some modeling challenges: (i) interfaces could be very thin layers that require the deployment of high-resolution model, (ii) interface mechanical properties are difficult to access given the well logs limitation in detecting them.

Aimene et al. [25] introduced the combined use of Anisotropic Damage Mechanics (ADaM) model and interface models in MPM to model the effects of interfaces in 2D and 3D hydraulic fractures problems. Multiple interface modeling tools were deployed [25] to achieve a better understanding of the impact of the interface properties on the fracture propagation initiation, growth and path. These tools include the Coulomb frictional contact and imperfect interface models. In the Coulomb frictional contact model, the contact between materials is modeled by setting the tangential traction S proportional to the normal force N at the interface by using the friction coefficient μ, i.e. $\mathbf{S} = \min(\mu\ \mathbf{N}, \mathbf{S}_{stick})$, where $\mathbf{S}_{stick}$ is tangential traction required for the interfaces to stick with zero discontinuity. In other words, the materials stick until the tangential forces required by sticking exceeds the frictional term after which the interface is modeled by frictional sliding.

In the imperfect interface model, interfaces are represented implicitly by their overall phenomenological effects described with a much smaller number of interfacial parameters [26]. In this context of imperfect interface model, surface drilling derived logs provide once more the information needed to detect major interfaces that could be responsible for loss of energy during hydraulic fracturing. These major interfaces could be inferred from the resulting Young's Modulus or/and Poisson's ratio computed from the surface drilling data and the CMSE. For example, **Figure 2C** shows an example of a Poisson's ratio pseudo-log estimated from surface drilling data where multiple spikes of high values indicate changes of lithology occurring where the well crosses a geologic interface. To illustrate the impact of interfaces and natural fractures on the fracture height, we will consider a 2D and a 3D case.

7.1 2D effect of weak interface on hydraulic fracture height

A specimen, under high vertical overburden stress, made of three layers alternating soft and stiff rock as estimated from the surface drilling estimation of Poisson's ratio and Young's Modulus is used to highlight the potential of a hydraulic fracture to develop a step-over behavior. We consider two cases: case 1 has perfect interfaces and case 2 has weak interfaces. To illustrate the step-over phenomenon, the two cases include a flaw (small fracture) located at the interface, close to the vertical path from the injection point (**Figure 11**).

Figure 11 shows that for case 1 with a strong interface, the vertical fracture growth was insensitive to the presence of the flaw and the fracture propagated in the direction of the applied stress giving rise to symmetric fracture half-heights (**Figure 11**, top). However, for case 2 with the weak interface and a flaw near the injection point, the hydraulic fracture was first arrested by the weak interfaces, and then stepping over occurred when the fracture reached the interface (**Figure 11**, bottom). The fracture height is much smaller in the case 2 with weak interfaces and a strong asymmetric fracture height was developed as a result of high shear in the weaker interfaces.

The weak interfaces gave rise to an asymmetric fracture half-height, where the flaw promoted the propagation toward the direction of its location. These flaws, which are mainly bed-bounded natural fractures, could affect the propagation path of the hydraulic fracture and generate asymmetric hydraulic fracture half-heights or arrest the parent fracture and promote the secondary child fracture. It is this complex geologic reality that makes current hydraulic fracturing simulators inadequate to capture the propped frac height. This geologic complexity gets more complicated as we consider actual 3D situations.

7.2 3D effects of dipping fractures planes on fracture geometry

There are multiple field conditions that cannot be modeled with 2D approximations, and full 3D modeling is needed. This is the case in a strike-slip stress regime, when dealing with natural fracture planes that are not vertical, or when stimulating with helical perforation, etc. In these cases, full 3D modeling tools are essential to accurately reproduce the 3D fracture propagation mechanisms that will lead to the correct fracture height. To illustrate a full 3D analysis, a 3D laboratory test in [27] was considered. Briefly, the laboratory specimen was made with a real reservoir sandstone. The specimen, subjected to the strike-slip stress regime, contains 3

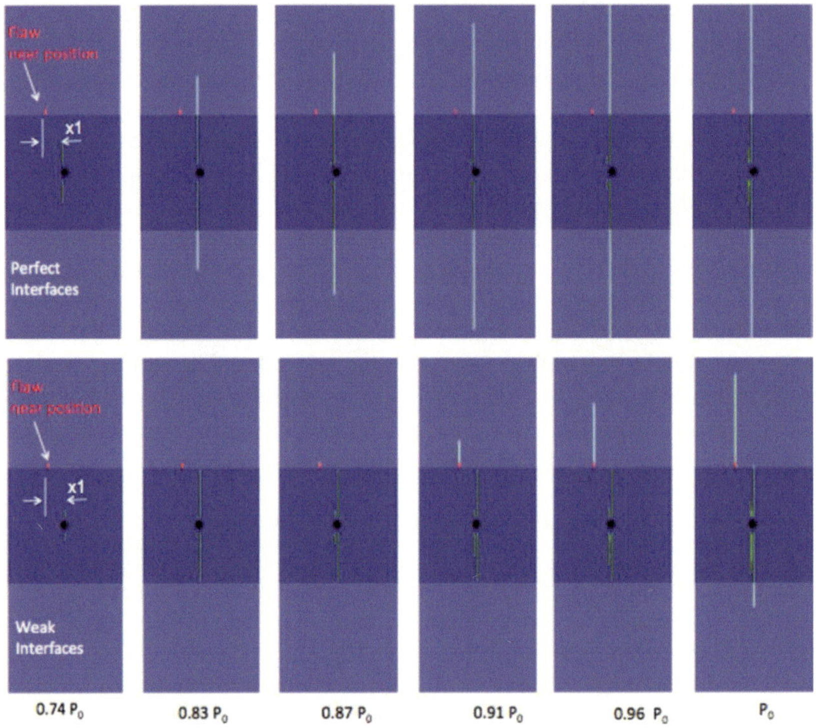

Figure 11.
Fracture propagation in the presence of a natural fracture flaw (in red) close to the vertical path from the injection point in specimen-2. Stiff layer (dark blue) bounded by soft layers (light blue) in the presence of a (top) perfect interfaces (case1, top), and weak interfaces (case 2, bottom). Notice the step-over occurred in the weak interface but not in the perfect interface. The animated figure can be seen in Video 1 available from (can be viewed at) https://youtu.be/oqDx96YXSvQ

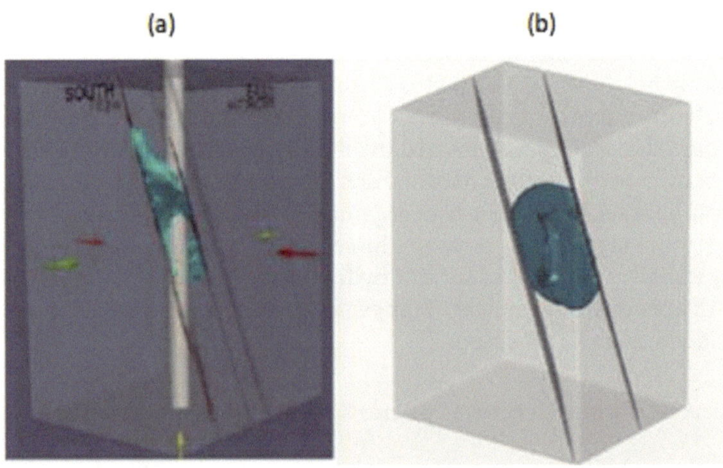

Figure 12.
(a) Experimental setup and laser scan showing final 3D fracture geometry (green) arrest against 2 natural fracture planes in [27] laboratory test (b) 3D ADaM MPM result showing the fracture geometry and its arrest against the dipping fracture planes represented as weakly bonded interface. Animated version of the 3D MPM result can be seen in Video 2 available from (can be viewed at) https://youtu.be/jdfDAM2qi-8.

natural fractures with a 40° strike relative to S_{Hmax} and 75° dip. In the experiment, the pressurization was accomplished by eight perforations, four in each side, parallel to the strike of the fractures. The natural fractures are not mineralized, so they were modeled with the Coulomb frictional contact law with μ = 0.85 according to the laboratory experiment. **Figure 12** shows the experimental results vs. the 3D MPM numerical model. The geomechanical modeling tool was able to reproduce the main features, especially the turning of the hydraulic fracture and arresting of the fracture by the nearest natural fractures. No crossing of natural fracture was observed neither in the numerical results nor in the experiment. This result highlights the ability of the combined use of the fully 3D ADaM model with the interface modeling tools in capturing the interaction between hydraulic and natural fractures in field conditions requiring full 3D geomechanical modeling tools.

The results from these decoupled processes, that attempt to quantify the geomechanical impact of geologic characteristics, can be used to constrain a 3D grid based planar hydraulic fracturing simulator that will include proppant transport.

8. Constrained 3D grid-based planar hydraulic fracturing simulator

Various hydraulic fracturing design scenarios can be analyzed starting from the geometric and engineered designs extracted in Section 3. The 3D reservoir models generated in Section 4 are fed in the hydraulic fracturing design model along with the various stress gradients in the lateral and vertical direction extracted from the differential stress model in Section 6. In the presence of stress gradients created by natural fractures, variable geomechanical properties and depleted reservoirs, the characteristics of the formation would be different on either side of the wellbore. This asymmetry and its correct quantification are important for an optimal well spacing of unconventional reservoirs. Fischer et al. [28] proposed a hydraulic fracture model (Eq. (5)) that explains the relative length change in two opposite wings of the hydraulic fracture.

$$a_1(t) = \frac{p_0^{net}}{g}\left(1 - \sqrt{1 - 2\left(\frac{g}{p_0^{net}}\right)a_2(t) - \left(\frac{g}{p_0^{net}}\right)^2 a_2\,(t)^2}\right) \tag{5}$$

The model is based on the lateral change in stress that would result in preferential growth of the hydraulic fracture in the direction of the decreasing confining stress. However, the model only explains the relativity of the asymmetric behavior in presence of the stress gradient (g) and does not specify the absolute length or width of the fracture. The relationship between the shorter wing a_1 (t) and longer wing a_2 (t) is given by Eq. (5) where p_0^{net} is the initial net pressure.

Using this concept, a 3D grid based planar hydraulic fracture model that includes proppant transport is developed. The unique feature of the hydraulic fracturing model is that it combines both analytical and numerical formulations. The effects of stress field change on the relative growth of the fracture is estimated by iterating for the optimum fracture height based on the amount of proppant available. The ability to develop a semi-analytical asymmetric fracture model that solves for the optimum fracture height and lengths is made possible by using the constraints of the geomechanical half lengths derived from the strain map and the estimated asymmetric fracture height derived from the ADaM geomechanical simulation.

The net pressure which is the difference between the fluid pressure and the minimum horizontal stress or the closure stress determines the initiation and propagation of a hydraulic fracture. The effect of stress gradient, along the fracture length is incorporated in the fundamental pressure balance equation at the fracture

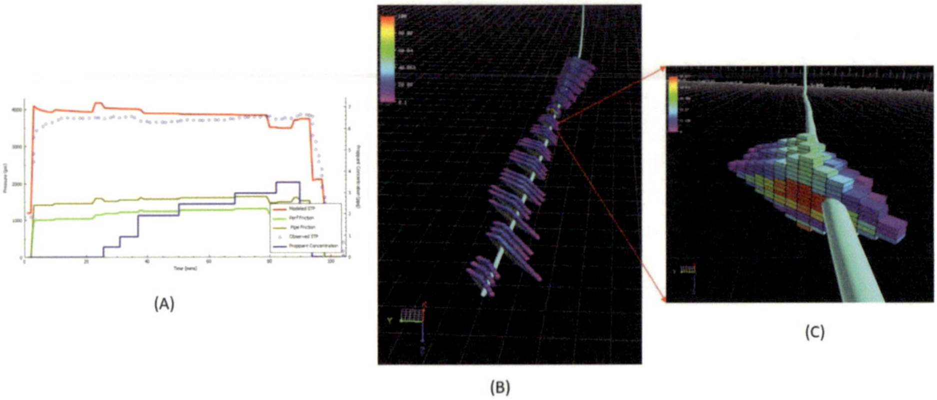

Figure 13.
(A) Pressure match at a stage and resulting (B) complex fracture geometry and conductivity along the wellbore with major lateral and vertical variations due to the variable nature of the rock properties captured by the surface drilling data along the well and at a single stage (C).

tip which determines the growth of the fracture. The fluid flow in the fracture is computed numerically. Using the relation of velocity of the fracture fluid and the fracture length, a time dependent solution is achieved which forms the basis of the semi-analytical model. All the rock properties and stresses are input in the hydraulic fracture simulator as 3D models as shown in **Figure 4**.

Using all these constraints as inputs in the 3D planar hydraulic fracture simulator, the pressure monitored during the actual fracturing treatment is easily matched (**Figure 13A**) by altering only the pipe and perf friction and the leak off coefficient which depends on the input porosity or natural fracture model. The resulting frac geometry at one stage (**Figure 13C**) or along the entire wellbore (**Figure 13B**) shows the major lateral and vertical dimension and conductivity variations owing to the variable nature of the rock properties captured by the surface drilling data. With this result at each well, we have all what is needed for the reservoir simulation which is needed to compute the Estimated Ultimate Recovery (EUR) and the resulting asymmetric depletion, a fundamental information needed for the planning of future wells.

9. Estimating well performances using reservoir simulation

For unconventional reservoirs undergoing hydraulic fracturing the long-term performance of a well could be represented by its Estimated Ultimate Recovery (EUR) which will also affect the extent of the depleted zone. The EUR and the size of the depleted zone will be major inputs needed for planning the development of an unconventional reservoir. This critical information requires the use of dynamic fluid flow reservoir simulation. In unconventional reservoir simulation, in addition to all the usual matrix properties required as input, the properties of the stimulated zone around the wellbore are necessary to capture the effects of the hydraulic fracturing and the development of the SRV. Multiple software tools and ways to provide the necessary input have been used. Despite major progress made recently in reservoir simulation, it remains a time-consuming task which prompted the need to develop a faster approach using the fast marching method (FMM). Both approaches are illustrated and compared with a field example.

9.1 Fast marching method (FMM)

Classical reservoir simulation using finite difference and finite volume have been widely used to simulate fluid flow in conventional oil and gas reservoirs. With the advent of unconventional reservoirs, these widely used classical reservoir simulation tools have been adapted to the particular nature of permeability generated through hydraulic fracturing. Current state of the art and the challenges associated with unconventional reservoirs are discussed in other books [29]. In this chapter we address the issue of computation time and the need for unconventional reservoir simulators to provide an estimate of the EUR and pressure depletion in the fastest possible way to enable the engineers and geoscientist to compare multiple development scenarios very quickly. In other words, given the nature of the unconventional process and its fast pace, how we can trade a reduction in accuracy for a much faster reservoir simulation?

Multiple efforts have been made to reduce computation time in reservoir simulators by replacing the flow equations with a proxy model based on neural networks [30] or response surfaces and experimental design [31]. Other efforts include the use of fast front-tracking techniques using streamlines [32] and, more recently, using a fast marching method or FMM [33]. We use the FMM in our approach of modeling unconventional reservoirs for the 3D estimation of pressure depletion.

The FMM is a front tracking algorithm. It has been applied to wave propagation, and medical imaging [34] problems. For the subsurface porous media flow, the pressure diffusivity equation can be simplified to an Eikonal equation and solved using the FMM to obtain the diffusive time of flight contours which is a proxy for the pressure depletion time [35]. The simulation is very fast compared to a finite difference simulator thus its valuable application to the fast-paced world of unconventional reservoirs. This speed is gained through a loss of accuracy which we could estimate by comparing the resulting pressure to those derived in a finite difference simulator.

9.2 Classical finite difference reservoir simulation

A dynamic model using a finite difference compositional reservoir simulator is built around a two well pad by using previously derived 3D properties (**Figure 4**) as well as the final hydraulic fracture geometry and its resulting conductivity (**Figure 13B**). Another derived key input is the inter-well permeability resulting from the interaction between the hydraulic and natural fractures and captured by the strain resulting from the geomechanical simulation using the MPM 2D plane strain framework. This strain is converted into an effective inter-well permeability (**Figure 14B**) using the approach shown in [36] where a calibration factor that relates strain to the matrix effective permeability is estimated through history matching. This calibration factor was very quickly estimated and a match was found for both wells A and B (**Figure 15**) for oil, gas and water rates using a bottom hole pressure. The major drawback of using a finite difference reservoir simulator is the intensive numerical computation required even in today's new generation parallel reservoir simulators. For example, the case described in this section required a 6-hour run on a good workstation. For unconventional reservoirs, we could trade the accuracy for a faster run time.

9.3 Unconventional reservoir simulation using fast marching method (FMM)

The motivation and the unique features of the Fast-Marching Method (FMM) simulator needed for unconventional reservoirs were described in Ouenes et al. [6] and Paryani et al. [18]. The input 3D models (**Figure 4**) and hydraulic fracture geometry (**Figure 14**) were input in the FMM simulator along with the PVT

(Pressure, Volume, Temperature) and other inputs. The resulting pressure depletion at the end of the simulation derived from the FMM simulator (**Figure 16B**) shows the same features as those seen in the pressure depletion estimated in the classical reservoir simulator (**Figure 16A**). The same conclusions can be seen when examining the pressure distribution in a cross-section view as shown in **Figure 17**.

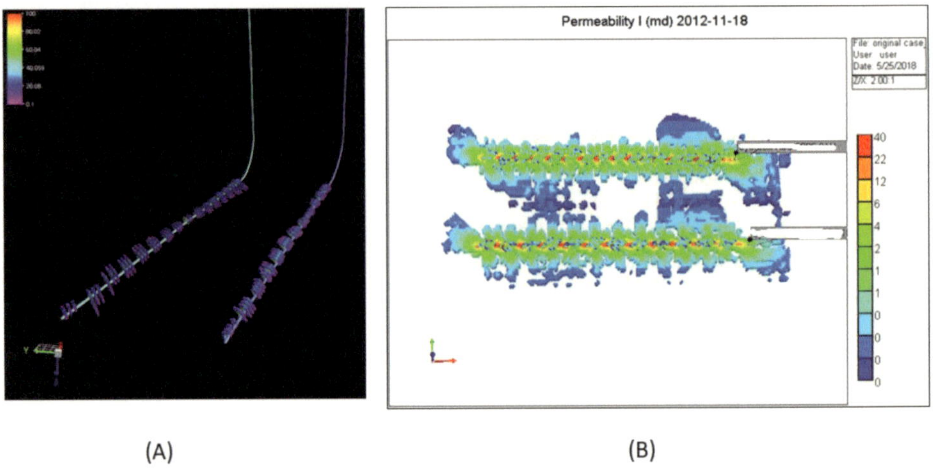

Figure 14.
(A) fracture geometry and conductivity resulting from the stimulation of two wells and (B) interwell permeability resulting from the interaction between the hydraulic and natural fractures.

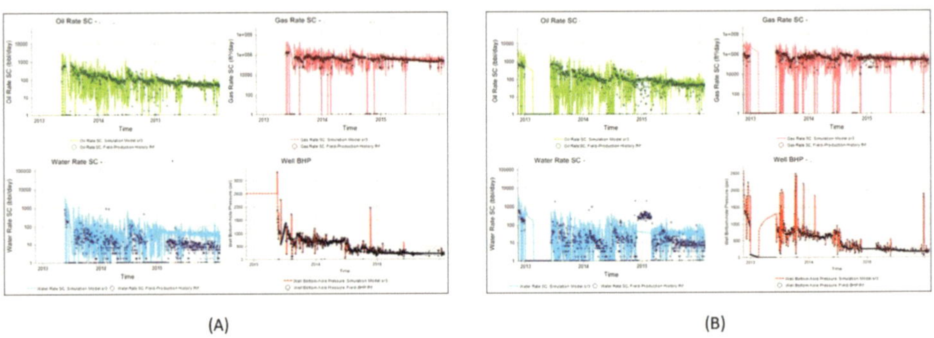

Figure 15.
History matching of oil (green), gas (red), and water (blue) by using the bottom hole pressure (BHP) as a constraint. Notice the good match of both well measurements A and B, achieved very easily and quickly by using one single history matching parameter.

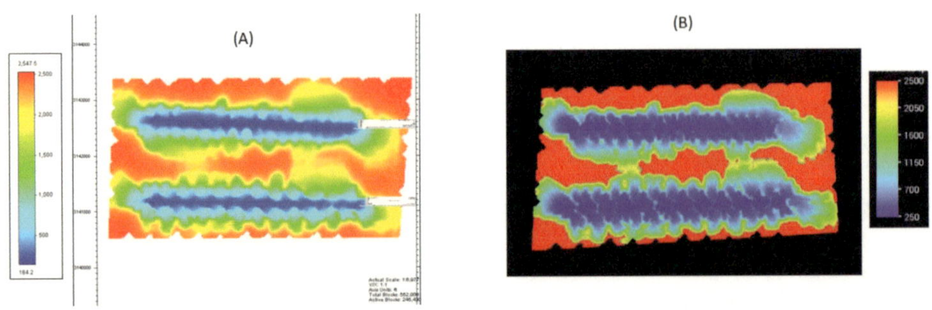

Figure 16.
(A) Areal view of the pressure depletion from finite difference reservoir simulator compared to the (B) pressure derived from the fast marching method simulator.

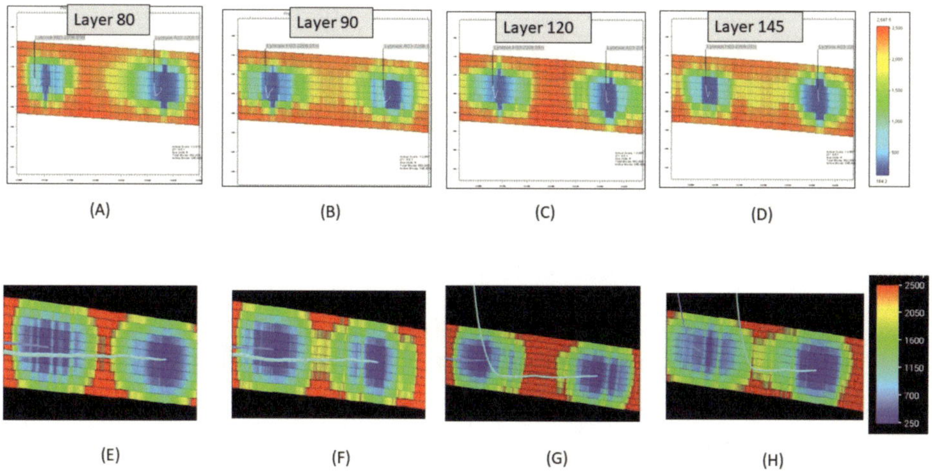

Figure 17.
(A)–(D) Cross section view of the pressure depletion from a finite difference reservoir simulator compared to the (E)–(H) pressure derived from the fast marching method (FMM) simulator.

There is a major difference between the classical finite difference reservoir simulation run and the one using the FMM: using the same computer, the full-scale heterogeneous model using the compositional finite difference reservoir simulator requires six (6) hours run time due to the large number of components while the FMM multi-phase black oil simulator results were derived in less than 1 minute.

With such a rapid evaluation tool and robust workflow that leverages the multiple constraints derived from the use of surface drilling data, the complex balance between finding the optimal Net Present Value (NPV) per well or per section could be easily estimated in few days or even hours. Using the current industry tools to achieve the same objective will take many weeks if not months and will have a large uncertainty if no well logs or seismic are available as it is very usual the case in unconventional reservoirs where well data and seismic are sacrificed at the altar of cost cutting measures. Fortunately, the surface drilling data provides a reasonable alternative that enables the entire reservoir modeling and management workflow.

10. Conclusions

The use of surface drilling data provides a reasonable engineering solution to the lack of well data in unconventional reservoirs. The correction of the MSE by removing friction losses turns surface drilling data into a major source of information for unconventional well planning. This information includes an estimation of geomechanical logs, pore pressure, stresses, porosity and natural fracture. These rock properties could be used as a first approximation in a well-centric approach to geoengineer completions. Moreover, combining these various logs from different wells into 3D reservoir models provides even more opportunities including using them in reservoir geomechanics, 3D planar hydraulic fracture design and reservoir simulation. The use of MPM modeling tools with the 2D horizontal plane strain framework allows the characterization of the lateral stress gradients, differential stress and the orientation of the maximum stress, all of which are key inputs needed to plan the optimal position and orientation of unconventional wellbores. Using geomechanical logs derived from surface drilling data to identify the geologic interfaces and deploying MPM tools in vertical 2D sections of the reservoirs, allows for capturing the effects of laminations and the loss of hydraulic fracturing energy in weak interfaces. These 2D decoupled approaches provide useful information to constrain a 3D grid based planar hydraulic fracturing approach. These geomechanical constraints define both the lateral and vertical directions which enable the estimation of distribution of proppants with a higher degree of certainty. The resulting realistic fracture geometry and its conductivity can be used in a commercial finite difference reservoir simulator or alternatively in a simplified fast marching method simulator providing similar information as the one given by the finite difference reservoir simulator, but in a fraction of the time. When using all these 3D models and their results in a fast marching method simulator the impact of the interference between wells and other optimization challenges can be estimated quickly while providing similar results as those derived with a finite difference reservoir simulator. By integrating the surface drilling data with 3D reservoir models, hydraulic fracturing design and reservoir simulation into a single software platform, this fast and constrained approach allows for a better management of unconventional wells within a competitive calculation time.

Acknowledgements

The authors would like to thank FracGeo and GO GeoEngineering staff for their contribution to the development, testing and deployment of these technologies. The contribution of Ron Dirksen, John Nairn, Djamel Sia, Bhavina Mistry, Mohcene Laraba, Thibaud Alengry, Nicholas Umholtz, Radouan Smaoui, Srichand Poludasu, Moussa Bari, Ravi Khandelwal, Xiaopeng Li, Bilel Attia, Aishwarya Gogoi, Abassi Ahmed, Dania Almadhun, Sonia K. Sanchez Lohff, Peter O'Conor, and Mark Morford to this large multi-disciplinary effort made it all possible. The authors also thank the Computer Modeling Group for providing the finite difference reservoir simulator.

Nomenclature

WOB	weight on bit,
D	hole diameter
T	torque
N	rotational speed
ROP	rate of penetration
N	rotational speed of the drill pipe
Kn	mud motor speed to flow ratio
Q	total mud flow rate
g	stress gradient
Po	net pressure which is the difference between the fluid pressure and the minimum or the closure stress
a_1 (t)	short wing in planar 3D fracture
a_2 (t)	long wing in planar 3D fracture
S_{Hmax}	maximum horizontal stress S_{Hmax}

Author details

Ahmed Ouenes*, Mohit Paryani, Yamina Aimene, Chad Hammerquist and Aissa Bachir
FracGeo, The Woodlands, Texas, USA

*Address all correspondence to: aouenes@gmail.com

References

[1] Rassenfos S. Real fractured rock is so complex it's time for new fracturing models. Journal of Petroleum Technology. 2018. https://www.spe.org/en/jpt/jpt-article-detail/?art=4645

[2] Ouenes A, Dirksen R, Paryani M, Rehman S, Bari M. Completion optimization while drilling: Geomechanical steering towards Fracable rock for optimal selection of stage spacing and cluster density in unconventional Wells. Society of Petroleum Engineers. 2017. DOI: 10.2118/188087-MS. https://www.onepetro.org/conference-paper/SPE-188087-MS

[3] Teale R. The concept of specific energy in rock drilling. International Journal of Rock Mechanics and Mining Sciences & Geomechanics, (Abstracts). 1965;**2**(1):57-73. ISSN 0148-9062

[4] Jacques A, Ouenes A, Dirksen R, Paryani M, Rehman S, Bari M. Completion optimization while drilling—Geomechanical steering towards Fracable rock using corrected mechanical specific energy. In: Unconventional Resources Technology Conference. 2017. DOI: 10.15530/URTEC-2017-2693870

[5] Paryani M, Sia D, Mistry B, Fairchild D, Ouenes A. Real-time completion optimization of fracture treatment using commonly available surface drilling and fracing data. Society of Petroleum Engineers. 2018. doi:10.2118/189810-MS. https://www.onepetro.org/conference-paper/SPE-190070-MS

[6] Ouenes A, Almadhun D, Oza S, Li SR, Smaoui R. Using surface drilling data for a geologically and geomechanically constrained 3D planar frac simulator and fast reservoir simulation—Application to engineered completions and well interference prevention. First Break. 2018;**36**(7):51-56

[7] Majidi R, Albertin M, Last N. Method for pore pressure estimation using mechanical specific energy and drilling efficiency. Society of Petroleum Engineers. 2016. DOI: 10.2118/178842-MS. https://www.onepetro.org/conference-paper/SPE-178842-MS

[8] Ouenes A. Practical application of fuzzy logic and neural networks to fractured reservoir characterization. Computer and Geosciences. 2000;**26**:953-962

[9] Jenkins C, Ouenes A, Zellou A, Wingard J. Quantifying and predicting naturally fractured reservoir behavior with continuous fracture models. AAPG Bulletin. 2009;**93**(11):1597-1608

[10] Warpinski NR, Teufel LW. Influence of geologic discontinuities on hydraulic fracture propagation. Journal of Petroleum Technology. 1987;**39**(02):209-220

[11] Aimene Y, Ouenes A. Geomechanical modeling of hydraulic fractures interacting with natural fractures—Validation with microseismic data and tracer data from the Marcellus and Eagle ford. Interpretation. 2015;**3**(3):1-18

[12] McKetta S, Vargas-Silva S, Ouenes A. Improved frac design and reservoir simulation using strain derived from geomechanical modeling—Application to the Fayetteville shale. Society of Petroleum Engineers. 2016. DOI: 10.2118/179157-MS. https://www.onepetro.org/conference-paper/SPE-179157-MS

[13] Ouenes A, Umholtz N, Aimene Y. Using geomechanical modeling to quantify the impact of natural fractures on well performance and

microseismicity: Application to the Wolfcamp, Permian Basin, Reagan County, Texas. Interpretation. 2016;**4**(2):SE1-SE15. https://doi.org/10.1190/INT-2015-0134.1

[14] Paryani M, Smaoui R, Poludasu S, Attia B, Umholtz N, Ahmed I, et al. Adaptive fracturing to avoid frac hits and interference: A Wolfcamp shale case study. Society of Petroleum Engineers. 2017. DOI: 10.2118/185044-MS. https://www.onepetro.org/conference-paper/SPE-187535-MS

[15] Ouenes A, Bachir A, Khodabakhshnejad A, Aimene A. Geomechanical modelling using poro-elasticity to prevent frac hits and well interferences. First Break. 2017;**35**(7):71-76

[16] Ouenes A, Hartley LJ. Integrated fractured reservoir modeling using both discrete and continuum approaches. Society of Petroleum Engineers. 2000. DOI: 10.2118/62939-MS. https://www.onepetro.org/conference-paper/SPE-62939-MS

[17] Ahmed Elfeel M, Jamal S, Enemanna C, Arnold D, Geiger S. Effect of DFN upscaling on history matching and prediction of naturally fractured reservoirs. Society of Petroleum Engineers. 2013. DOI: 10.2118/164838-MS. https://www.onepetro.org/conference-paper/SPE-164838-MS

[18] Paryani M, Bachir A, Smaoui R, Umholtz N, Khodabakhshnejad A, Ouenes A. Engineered completion and well spacing optimization using a geologically and geomechanically constrained 3D planar frac simulator and fast marching method: Application to eagle ford. Society of Petroleum Engineers. 2018. DOI: 10.2118/190070-MS. https://www.onepetro.org/conference-paper/SPE-190070-MS

[19] Ruppel SC, Jones RH, Breton CL, Kane JA. Preparation of maps depicting geothermal gradient and Precambrian structure in the Permian Basin: The University of Texas at Austin, Bureau of Economic Geology. Contract Report to the U.S. Geological Survey. 2005, order no. 04CRSA0834 and requisition no. 04CRPR01474

[20] Stoeser DB, Green GN, Morath LC, Heran WD, Wilson AB, Moore DW, Van Gosen S. Preliminary integrated geologic map databases for the United States central states: Montana, Wyoming, Colorado, New Mexico, North Dakota, South Dakota, Nebraska, Kansas, Oklahoma, Texas, Iowa, Missouri, Arkansas, and Louisiana Version 1.2, U.S. Geologic Survey. 2007. Open-File Report 2005-1352

[21] Ptaszynska A, Green S, Hupka J, Suarez-Rivera R. Mineral and organic matter constituents in weak interfaces in shales. In: 50th US Rock Mechanics Geomechanics Symposium, Houston, Texas. 2016

[22] Sonnenfeld M, Ohlson C, Zahm C, Odegard M. The impact of multiple, thin bentonites on proppant placement and effective fracture continuity within the niobrara formation, Weld County, Colorado. Unconventional Resources Technology Conference. 2015. DOI: 10.15530/URTEC-2015-2171548

[23] Smart JK, Goodluck LO, Morris AP, McGinnis RN, Ferril DA. Geomechanical modeling of hydraulic fracturing: Why mechanical stratigraphy, stress and pre-existing structure matter. AAPG Journal. 2014;**98**(11):2237-2261

[24] Cohen C-E, Kresse O, Weng X. Stacked height model to improve fracture height growth prediction and simulate interactions with multi-layer DFNs and ledges at weak zone interfaces. Society of Petroleum Engineers. 2017. DOI: 10.2118/184876-MS. https://www.onepetro.org/conference-paper/SPE-184876-MS

[25] Aimene Y, Hammerquist C, Nairn JA, Ouenes A. 3D anisotropic damage mechanics for modeling interaction between hydraulic and natural fracture planes in a layered rock-application to eagle ford and Wolfcamp. Unconventional Resources Technology Conference. 2018. DOI: 10.15530/URTEC-2018-2902985

[26] Hashin Z. Thermoelastic properties of fiber composites with imperfect interface. Mechanics of Materials. 1990;**8**:333-348

[27] Yang X, Burghardt J, Zhang H, Zhang Y, Zhang F, Pei J, et al. Experimental study of hydraulic fracture/natural fracture interaction on a tight sandstone formation. In: Unconventional Resources Technology Conference. 2016. DOI: 10.15530/URTEC-2016-2460449

[28] Fischer T, Hainzl S, Dahm T. The creation of an asymmetric hydraulic fracture as a result of driving stress gradients. Geophysical Journal International. 2009;**179**:634-639. DOI: 10.1111/j.1365-246X.2009.04316.x

[29] Islam MR, Hossain ME, Moussavizadegan SH, Mustafiz S, Abou-Kassem JH. Front matter. In: Islam MR, Hossain ME, Moussavizadegan SH, Mustafiz S, Abou-Kassem JH, editors. Advanced Petroleum Reservoir Simulation. 2016. DOI: 10.1002/9781119038573.fmatter. https://onlinelibrary.wiley.com/doi/abs/10.1002/9781119038573.fmatter

[30] Ouenes A, Doddi R, Lin Y, Cunningham G, Saad N. A new approach combining neural networks and simulated abnnealing for solving petroleum inverse problems. In: Proceedings of the 4th European Conference on the Mathematics of Oil Recovery [ECMOR-IV '94], Roros, Norway. 1994

[31] Eidi AL, Holden L, Reiso E, Aanonsen SI. Automatic history matching by use of response surfaces and experimental design. In: Proceedings of the 4th European Conference on the Mathematics of Oil Recovery [ECMOR-IV '94], Roros, Norway. 1994

[32] Barman I, Ouenes A, Wang M. Fractured reservoir characterization using streamline-based inverse modeling and artificial intelligence tools. Society of Petroleum Engineers. DOI: 10.2118/63067-MS. https://www.onepetro.org/conference-paper/SPE-63067-MS

[33] Lino A, Vyas A, Huang J, Datta-Gupta A, Fujita Y, Bansal N, et al. Efficient modeling and history matching of shale oil reservoirs using the fast marching method: Field application and validation. Society of Petroleum Engineers. 2017. DOI: 10.2118/185719-MS. https://www.onepetro.org/conference-paper/SPE-185719-MS

[34] Sethian JA, Popovici AM. 3-D traveltime computation using the fast marching method. Geophysics. 1999;**64**(2):516-523. DOI: 10.1190/1.1444558

[35] Datta-Gupta A, Xie J, Gupta N, King MJ, Lee WJ. Radius of investigation and its generalization to unconventional reservoirs. Society of Petroleum Engineers. 2011. DOI: 10.2118/0711-0052-JPT. https://www.onepetro.org/journal-paper/SPE-0711-0052-JPT

[36] Ahmed I, Oza S, Vargas-Silva S, Poludasu S, Paryani M, Attia B, et al. Cracking the code of unconventional reservoir simulation. Society of Petroleum Engineers. 2017. DOI: 10.2118/188111-MS. https://www.onepetro.org/conference-paper/SPE-188111-MS

Chapter 6

Elastic-Based Brittleness Estimation from Seismic Inversion

Maman Hermana, Deva Prasad Ghosh and Chow Weng Sum

Abstract

Information about mechanical rock properties is essential when tight reservoir is to be stimulated using hydrofracturing technique. The brittle area has to be considered as a priority region for determining the location of hydrofracturing initiation. Seismic data are commonly used to estimate the geomechanical properties such as brittleness average from elastic properties: Poisson's ratio and Young's modulus. This paper discusses the process of brittleness estimation based on elastic properties, which can be derived by inverting the pre-stack seismic data that can produce acoustic impedance, shear impedance, and density simultaneously. Novel methods, scaled inverse quality factor of P-wave (SQp) and scaled inverse quality factor of S-wave (SQs) attributes, have been used for identification of brittleness, fracture density, and hydrocarbon bearing in the fractured basement reservoir. The effectiveness of the proposed method has been tested in the field, which is consistent with fracture density log from formation micro-imager (FMI) log and hydrocarbon column data. The result showed that there is a significant correlation between brittleness, estimated from elastic properties, and fracture density logs. New attributes, the SQp attribute is potentially to be used as a fracture density indicator, while SQs attribute indicates the existence of hydrocarbon, which is confirmed with neutron porosity-density logs.

Keywords: brittleness average, attribute, crack density, fractured basement, elastic properties

1. Introduction

Fractures are important for improving permeability in unconventional reservoirs including shale gas, coal bed methane, tight gas sand, and fractured basement reservoir. In these reservoirs, hydrofracturing is commonly practiced to stimulate fractures and to significantly improve oil/gas flow. The target of hydrofracturing technique is focused on the brittle area (the area with a high tendency to break), which is expected to be able to generate more fractures. To support this objective, understanding of mechanical properties (such as brittleness) of the rock is very important. Estimation of brittleness from seismic data is an important task for better well hydrofracturing and drilling placement.

The success of hydrofracturing depends on the geomechanical brittleness of the formation; brittle rocks tend to generate more fractures compared to ductile rocks. Brittleness is the measurement of stored energy before failure and is a function of rock strength, lithology, texture, effective stress, temperature, fluid type, digenesis, and TOC [1].

The brittleness is determined by a number of mineral contents of rock. The most brittle minerals like quartz and the less percentage of ductile minerals like clay mineral in the rock tend to make rock more brittle. Rock physics shows that the mineral content determine the elastic properties of rock. Hence, it is reasonable to estimate brittleness using elastic properties. However, the selection of which elastic properties can be used to indicate brittleness is the main task in seismic quantitative interpretation. Estimation of brittleness index which is based on calculation of mineral content and brittleness average which is based on elastic properties and can be derived from seismic data has been successfully applied in the shale gas field [2].

During unconventional reservoir exploration and development, not only how to find out the most brittle area where expected fracture can be generated during horizontal well drilling and hydrofracturing but also how to find the sweet spot where the hydrocarbon is accumulated largely and also how to estimate the capacity and reserve of unconventional reservoir are very challenging. As in case of fractured basement reservoir, the problem on how to find the possible location of generated secondary porosity and permeability and to find where the hydrocarbon accumulation is and what type of hydrocarbon is there is still difficult to be solved and needs advance tool to make more accurate and significant during quantification.

The main objective of this paper is to introduce a new workflow for unconventional reservoir characterization by introducing new attributes: scaled inverse quality factor of P-wave (SQp) and scaled inverse quality factor of S-wave (SQs). These attributes are derived from the attenuation concept through rock physics approximation, which can be implemented on the result of seismic inversion. The existing method, brittleness average, is commonly used to indicate the brittle rock calculated from elastic properties; Poisson's ratio and Young's modulus will be discussed and compared with new attributes to indicate the fracture density. A well data example from fractured basement reservoir in the Malaysian Basin is used to test the performance of these methods to indicate fracture density and hydrocarbon column which also will be discussed.

2. Brittleness index (BI) and brittleness average (BA)

Brittleness of rock has been defined in different ways. Jarvie [2] defines the brittleness index (BI) as a fraction of the mineral composition of rock, while Grieser and Bray [3] define brittleness average (BA) as purely related to the elastic properties of the rock.

As mineral composition of rock defines its brittleness, the number of fractions of most brittle mineral impacts on the rock brittleness. Brittleness index (BI) is formulated as

$$BI_{Jarvie} = \frac{Qz}{Qz + Ca + Cly} \tag{1}$$

where Qz, Ca, and Cly are the fractional quartz content, calcite content, and clay content, respectively.

For wells that are located where the composition of mineral can be properly determined, the BI can be calculated. However, away from the well, the BI is difficult to be estimated due to the difficulties in predicting the mineral content distribution. Hence, it is still difficult to use this technique to estimate brittleness three-dimensionally, because of the challenge in estimating mineral content from seismic data.

Grieser and Bray [3] proposed the use of brittleness average (BA) to express the brittleness of the rock. Brittleness average is calculated based on elastic properties, i.e., normalized Poisson's ratio and Young's modulus. By using this relation, estimation of brittleness in a wider area is possible. Both Young's modulus and Poisson's ratio can be derived from seismic data through seismic inversion. Hence, using this technique the brittleness of rock in terms of BA can be estimated from seismic data.

Young's modulus (E), representing the stiffness of the rock, can be defined in terms of bulk modulus (κ) and Poisson's ratio (σ) as.

$$E = -3\,\kappa\,(1\text{–}2\,\sigma) \tag{2}$$

On the other hand, Poisson's ratio can be derived from P-wave (Vp) and S-wave (Vs) velocities:

$$\sigma = \frac{Vp^2 - Vs^2}{2Vp^2 - 2Vs^2} \tag{3}$$

By substituting Eq. (3) in Eq. (2), the Young modulus is expressed as

$$E = \rho Vs^2 \frac{(3Vp^2 - 4Vs^2)}{Vp^2 - Vs^2} \tag{4}$$

Hence, the brittleness average (BA) is expressed in Rick's relation [1]:

$$BA = \frac{1}{2}\left(\frac{E - E_{min}}{E_{max} - E_{min}} + \frac{PR - PR_{max}}{PR_{min} - PR_{max}}\right) \times 100 \tag{5}$$

where E_{min} and E_{max} are the minimum and maximum Young's modulus and PR_{min} and PR_{max} are the minimum and maximum Poisson's ratio.

To evaluate the correlation between brittleness average and brittleness geomechanically, the brittleness average is tested against well logs of the domain and compared to geomechanical properties obtained from the formation micro-imager (FMI) log (**Figures 1** and **2**). Examples are taken from a fractured basement reservoir field located in Malaysian offshore. This field is located at a margin of the basin as permo-carboniferous metasediments and volcanic, cretaceous granites, or possibly cretaceous rift fill and mesozoic to carboniferous carbonates and mesozoic granites [4]. The lithology of the offshore basement for this area is described in Tjia et al. [5] based on well drilling distribution with pre-tertiary rock penetration.

Figure 1 shows the brittleness average logs which is calculated from normalized Poisson's ratio and Young's modulus logs and compared to the core of two different depth samples. The first sample (upper right) was taken from the depth where the brittleness average value is low. In this depth, the core sample showed that only a view number of fractures appear. The crack density of this core sample is low which is correlated with a low value of brittleness average. The second sample (bottom right) was taken from the depth where the brittleness average value is high. Many fractures appeared in this core sample. The core data is taken from a region where the rock is more brittle, which is also associated with a high value of brittleness average in the log. In other words, the intensity of fracture of the rock can be determined by the brittleness average log.

The feasibility study on well log data shows that the brittleness average has a good correlation with fracture density. A high fracture density area is associated

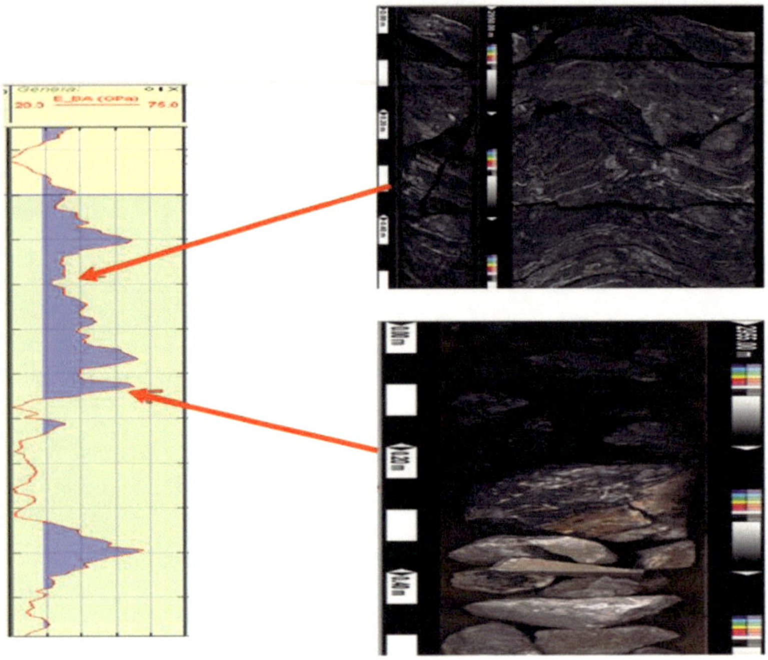

Figure 1.
Brittleness average log (left) compared to the FMI data (right). The high value of brittleness average associated with high crack density.

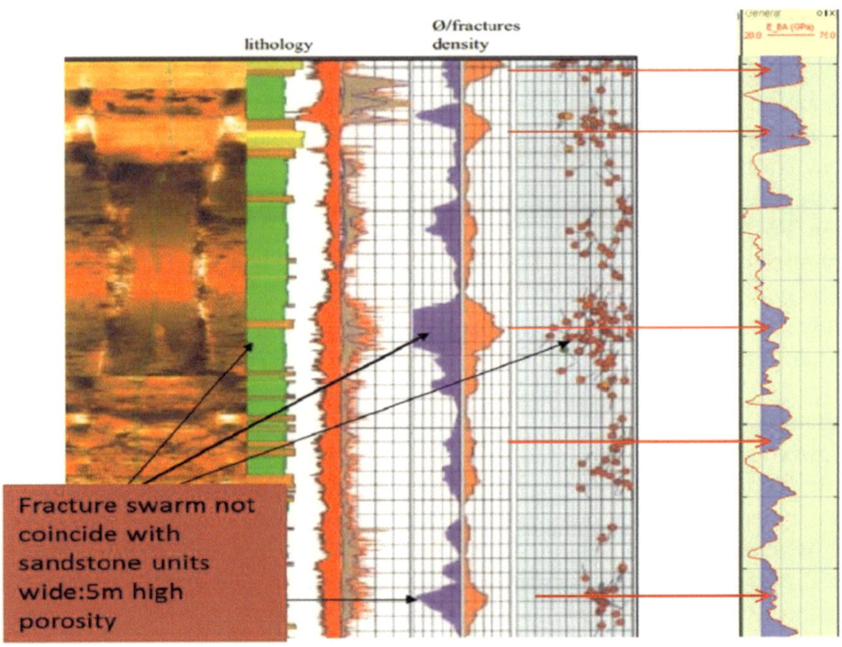

Figure 2.
Lithology and fracture density logs (left) compared with brittleness average (right log).

with high brittleness average log. Because of the elastic properties, Young's modulus and Poisson's ratio can be extracted from seismic data through inversion result; therefore, the brittleness average, which is associated with fracture density also can be calculated from inversion result.

3. SQp and SQs attributes

As in the viscoelastic medium, attenuation and phase velocity of plane wave propagation are governed by the Kramers-Kronig relations [6]. The maximum value of quality factor of P-wave and S-wave which represent the degree of attenuation can be estimated from basic elastic properties of compressional modulus (M) and shear modulus (G) at high- and low-frequency conditions:

$$\begin{aligned} 2Q_p^{-1} &= \frac{M_\infty - M_0}{\sqrt{M_0 M_\infty}} \\ 2Q_s^{-1} &= \frac{G_\infty - G_0}{\sqrt{G_0 G_\infty}} \end{aligned} \tag{6}$$

where the indexes (∞) and (0) represent relaxed and unrelaxed conditions, which are still difficult to be measured directly from seismic data.

A high- and low-frequency measurement, in rock physics, can be assumed as an effect of crack by the Hudson crack theory [6]. The changes of anisotropy stiffness component are associated with the difference between compressional modulus at high and low frequencies and can be correlated with Lame parameters: λ and μ. The change in bulk modulus is approximated by

$$\begin{aligned} M_\infty - M_o &= \Delta c_{11}^{Hudson} \\ &\approx \varepsilon \frac{\lambda^2}{\mu} \frac{4(\lambda + 2\mu)}{3(\lambda + \mu)} \end{aligned} \tag{7}$$

And the change in the shear modulus is approximated by

$$\begin{aligned} G_\infty - G_o &= \Delta c_{44}^{Hudson} \\ &\approx \varepsilon\mu \frac{16(\lambda + 2\mu)}{3(3\lambda + 4\mu)} \end{aligned} \tag{8}$$

where ε is the crack density, which is estimated from porosity and aspect ratio (α) as$\varepsilon = 3\phi/(4\pi\alpha)$. By assuming that $M = \sqrt{M_o M_\infty}$ and $G = \sqrt{G_o G_\infty}$, the Qp and Qs are formulated as [6]

$$\begin{aligned} Qp^{-1} &= \frac{2}{3}\varepsilon \frac{(M/G - 2)^2}{(M/G - 1)} \\ Qs^{-1} &= \frac{8}{3}\varepsilon \frac{(M/G)}{(3M/G - 2)} \end{aligned} \tag{9}$$

Information on crack density is indicated by Qp^{-1} and Qs^{-1}. If the crack density of the rock increases, the secondary porosity related to the crack will increase, while the bulk density decreases. In other words, an increase in crack density will be followed by a decrease in bulk density. Hence, Eq. (9) can be approximated as [7]

$$\begin{aligned} SQp^{-1} &\equiv \frac{5}{6}\frac{1}{\rho} \frac{(M/G - 2)^2}{(M/G - 1)} \\ SQs^{-1} &\equiv \frac{10}{3}\frac{1}{\rho} \frac{(M/G)}{(3M/G - 2)} \end{aligned} \tag{10}$$

where SQp^{-1} and SQs^{-1} are defined as **scaled inverse Qp (SQp)** and **scaled inverse Qs (SQs)**, which are indicating the attenuation of P and S wave, respectively. These parameters can be extracted from seismic data through inversion results where M/G is approximated from P- and S-wave velocity ratio or Vp/Vs.

4. Approximation of SQp and SQs attributes in the amplitude versus offset (AVO) domain

Derivations of SQp and SQs in the previous sub-chapter show that the parameters are taken from elastic properties extracted from seismic inversion results. To estimate the SQp and SQs directly from seismic data, the approximation of SQp and SQs through the AVO method is proposed.

To understand the approximation of SQp and SQs in the AVO domain, we started with the concept of AVO attributes: AVO intercept and AVO gradient methods. Castagna et al. [8] interprets the AVO intercept and AVO gradient using crossplot method (reader who is interested to get more details on this method can refer to Castagna et al. [8]). In AVO crossplot method, the diagonal line indicates the shale background, and potential hydrocarbon reservoirs will be identified as an AVO anomaly. Determination of the anomaly is measured from the shale background line (**Figure 3**).

The trend or background of Vp/Vs can be approximated with the formula

$$a_b = B/A \tag{11}$$

where B and A are AVO gradient and AVO intercept, respectively. Using Gardner equation [9], the relation between contrast density and contrast velocity is approximated with the formula

$$\frac{\Delta\rho}{\rho} \approx .25\frac{\Delta V_P}{V_P} \tag{12}$$

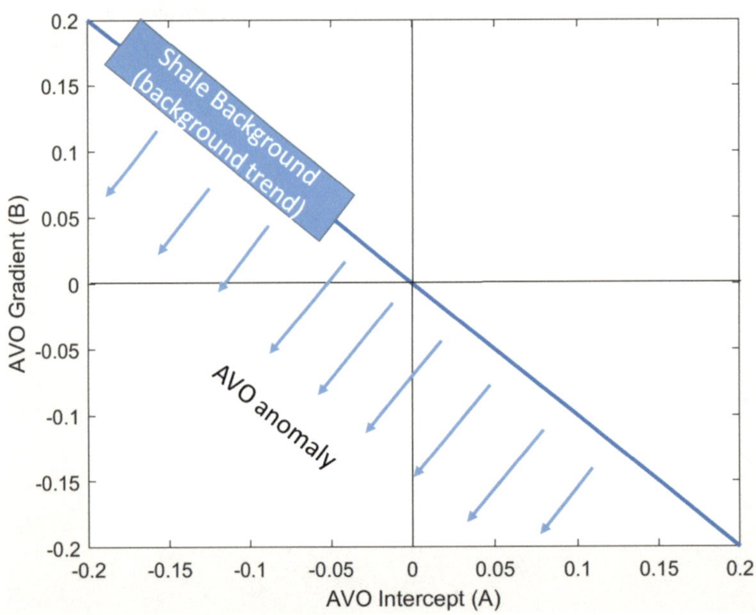

Figure 3.
AVO crossplot method. Diagonal is showing shale background, and hydrocarbon is identified as AVO anomaly.

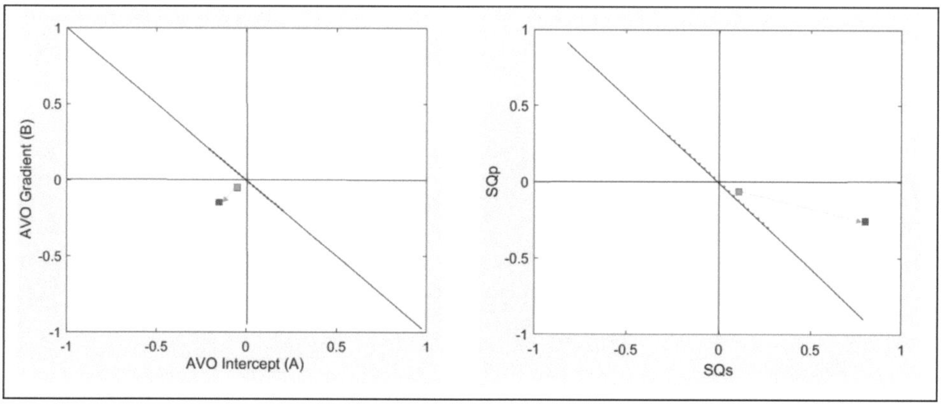

Figure 4.
Comparison between conventional AVO crossplot and new SQp-SQs crossplot.

where Δ represents the different properties between the upper and lower media. Hence, $\Delta\rho = \rho_2 - \rho_1$, and $\rho = \rho_2 + \rho_1$, $\Delta V_P = V_{P2} - V_{P1}$, and $V_P = V_{P2} + V_{P1}$. The velocity contrast is approximated by

$$\frac{\Delta V_P}{V_P} \approx (8/5)A \tag{13}$$

Combining Eqs. (12) and (13) and substituting Eq. (10), SQs^{-1} and SQp^{-1} can be approximated with

$$\begin{aligned} \mathrm{SQp} &\approx \left(\frac{1}{3}\right) A \left(\frac{(-3 + 2(\mathrm{A}+\mathrm{B}))^2}{2(1-(\mathrm{A}+\mathrm{B}))}\right) \\ \mathrm{SQs} &\approx \left(\frac{4}{3}\right) A \left(\frac{(2(\mathrm{A}+\mathrm{B})-1)}{2(\mathrm{A}+\mathrm{B})+1}\right) \end{aligned} \tag{14}$$

where A and B are the intercept and gradient attributes. Eq. (14) shows that the SQp and SQs attributes can be approximated from the intercept and gradient of AVO.

The crossplot of both methods can be illustrated in **Figure 4**. The potential of hydrocarbon reservoir from oil (blue) to gas (red) is plotted close to each other in the conventional AVO crossplot. However, that anomaly is boosted in the SQp-SQs crossplot. The two different potential reservoirs are separated significantly.

5. Numerical testing of SQp and SQs attributes on rock physics model

There are at least two main aspects recorded on seismic data: lithological effect and fluid effect. To understand the effect of both lithological and fluid changes, a rock physics model of soft sediment is used to test the response of SQp and SQs attributes. Initial parameters of model are derived from well log data. Gassmann's fluid substitution was applied on the model to get the new lithology and pore fill at different water saturations and porosities. In this test, the lithological effect is represented by the changes of porosity, while the fluid effect is represented with different water saturation conditions. Water saturation is set from 1 to 100%, where gas is used as a complement, and porosity changes are set from 7 to 25%, while the aspect ratio is assumed to be 0.1. The responses of SQp and SQs attributes at different water saturation and porosity changes are shown in the following figure.

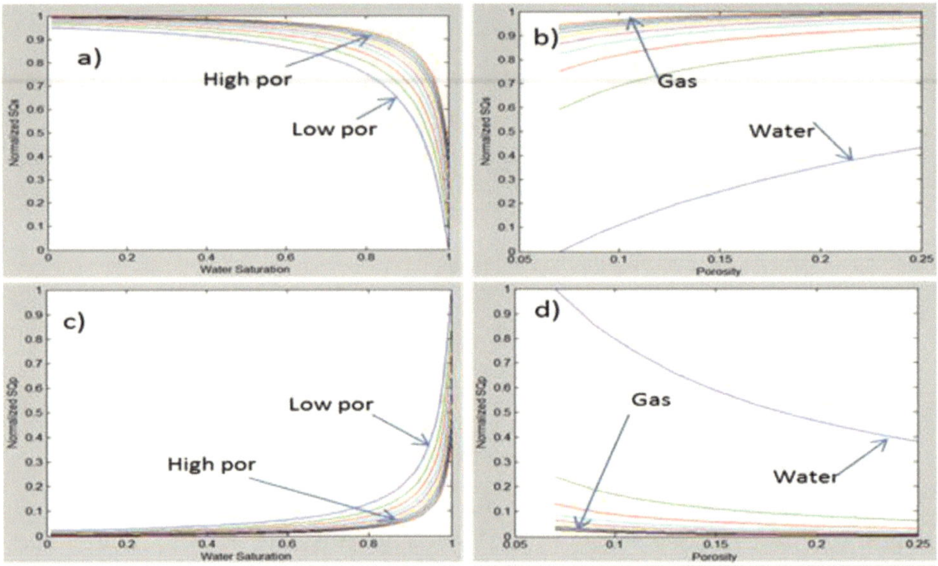

Figure 5.
Responses of SQp and SQs attributes on water saturation changes (left column) and porosity changes (right column). (a) SQs attribute versus water saturation, (b) SQs attribute versus porosity, (c) SQp attribute versus water saturation, and (d) SQp attribute versus porosity. The initial state of the models is Vp = 2231.9 m/s, Vs = 1127.04 m/s, density = 2.11 g/cc, Vsh = 0.56, Sw = 0.49, and porosity = 25%.

The responses of SQs attribute decrease when water saturation increases (**Figure 5a**). In constant porosity, the SQs value of gas sand is higher than water sand. The porosity changes also affect the SQs attribute; increasing in porosity is followed by increasing in SQs (**Figure 5b**). It can be interpreted also that when porosity of rock increases, the number of fluid inside the rock also increases, which tends to increase the SQs value. This attribute is sensitive to fluid changes (saturation) which means that SQs can be correlated to fluid content conditions. For every different conditions, gas sand has higher SQs value compared to water sand.

Figure 5c and **d** shows the SQp responses due to water saturation and porosity changes, respectively. The responses of SQp attribute increase when water saturation increases. However, the increment is significant when the water saturation is close to fully water-saturated conditions. Water saturation is from 0 to about 80%; the increment of SQp is not significant. In the condition where the gas saturation is low (where gas saturation is about 5 or 95% of water), SQp value increases exponentially (**Figure 5c**). This phenomenon is the same as in Gassmann's fluid substitution case where only 5% gas can boost seismic velocity exponentially. On the other side, when porosity increases (where the fluid content is more), the SQp values decrease (**Figure 5d**). It tells us that the number of fluid does not so much affect the SQp. In this example the change of lithology is represented by the change of porosity. SQp is more affected by lithology rather than fluid content. Hence, the SQp attribute might be better as a lithology indicator, while the SQs attribute would be better as a fluid indicator. This hypothesis will be proven by testing the attributes using real data.

6. SQp and SQs attribute responses on well domain

Numerical test of SQp and SQs attributes on rock physics shows that the SQp attribute is an indicator of lithological changes, while the SQs attribute can be used to denote fluid changes. The application of this concept on real well log data can be used to justify this assertion. To do so, another test is carried out to investigate the performance of these attributes in identifying the fluid type and lithological effect. Visual comparison and crossplotting of these attributes on well log data and comparisons with other lithology and fluid indicator (gamma ray and water saturation) logs are presented in **Figure 2**.

Figure 6a shows three different reservoir targets, two reservoirs are saturated by gas, and another reservoir is wet (water saturated). All three different reservoirs are indicated as low SQp values. In SQp attribute there is no different responses between gas sand and wet sand; all sand formations are shown as low SQp value. This example shows that SQp attribute is not sensitive with fluid type, only sensitive to the lithology changes. The formation of shale and sand is distinguished clearly as well as in the gamma ray log, while in terms of SQs attribute, both gas and

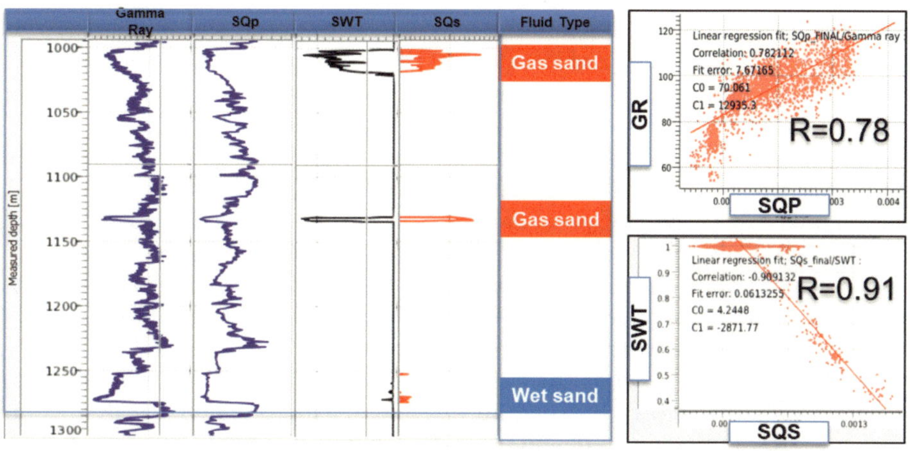

(a)

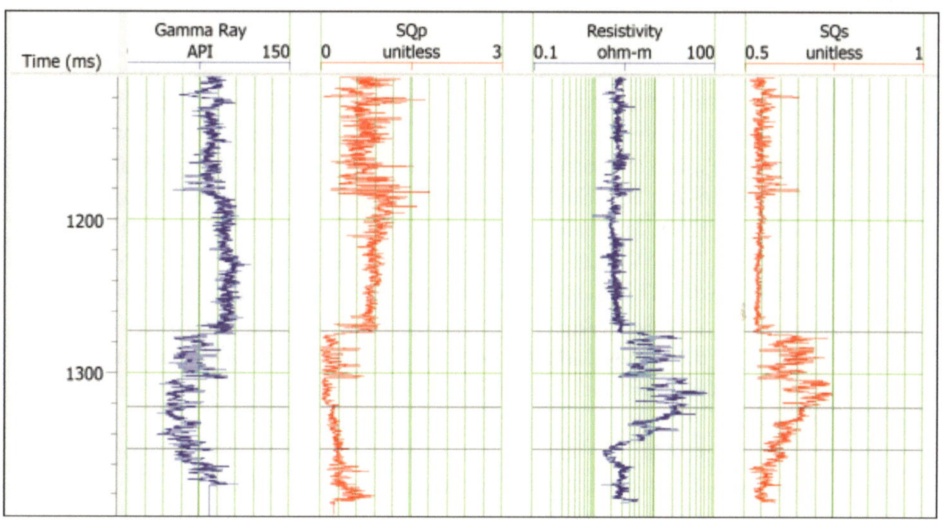

(b)

Figure 6.
(a) SQp and SQs responses compared to lithology log (gamma ray) and water saturation log and its coefficient correlations are obtained from the crossplot (a, right). (b) SQp and SQs test on different well. The SQp attribute is also similar to gamma ray log, and SQs is similar to resistivity logs.

sand reservoirs have higher value than wet sand. It shows that this attribute is more sensitive to the fluid type than lithology changes. The confirmation of the fluid content is shown by water saturation log, which is also similar to the SQs log.

An example from another field in offshore Malaysia (**Figure 6b**) shows that SQp response is also similar to the gamma ray logs, which supports the hypothesis that this attribute can be used to identify lithology changes in the same way as the gamma ray. Meanwhile, the SQs attribute, which was compared to the resistivity logs, shows that this attribute has high similarity to the resistivity logs. Resistivity log is commonly used to identify the fluid type of the formation; this log is sensitive with the changes of fluid type. Thus, SQs attribute also has potential to be used as fluid indicator. In **Figure 6b**, hydrocarbon formation is indicated by high resistivity value which also is shown in the SQs log. Hydrocarbon formation is indicated as a high SQs value, while brine/water sand will have a lower value.

The separation between lithology and fluid effect is identified easily in the crossplot. Optimum separation between lithological and fluid effects should be orthogonal to each other. To test the effectiveness of SQp and SQs attributes in discriminating the lithology and fluid effect, the crossplot of SQp-SQs has been compared with other elastic properties: lambda-rho vs. mu-rho crossplot as shown in **Figure** 7. The first and second Lame constants (see Section 3) multiplied with density are defined as lambda-rho and mu-rho, respectively. These attributes are a pair of elastic properties that are commonly used to discriminate the lithology and fluid. **Figure** 7 shows two different crossplots of mu-rho versus lambda-rho and SQp versus SQs crossplots color-coded by lithology log. All attributes, mu-rho, lambda-rho, SQp, and SQs, are calculated from the same sonic, shear, and density logs. The end members of lithology are classified into four types of lithology: shale sand, wet sand, shaly sand/siltstone sand, and pay sand. The types of lithology are defined by taking the cutoff on the volume of clay, gamma ray, porosity, and water saturation logs. The cutoff for shale was Vclay >0.4, gamma ray >80; shaly siltstone is Vclay <0.4, gamma ray <80, and porosity <0.05; wet sand is Vclay <0.4, porosity >0.05, water saturation >0.85; and pay sand is Vclay <0.4, porosity >0.05, and water saturation <0.85. The lithology log was used to identify the performance or sensitivity of attribute or elastic properties in predicting the lithology and hydrocarbon.

In the mu-rho versus lambda-rho crossplot (**Figure 7a**), gas sand still can be separated from wet sand and shale. However, in this crossplot, it is still difficult to define the separation between lithological and fluid effects. Conversely, in SQp-SQs crossplot (**Figure 7b**), it is not only gas sand and wet sand that are separated, but also the effect of lithology and fluid are separated optimally. In the SQp axis, different lithologies, shale and sand, are distinguished, while in SQs axis that lithology is not separated. The SQs can distinguish gas sand (net pay), wet sand, and shaly sand stone clearly. This SQs axis shows the effect of fluid. Therefore, SQp versus SQs shows an optimum separation between lithological and fluid effect. Lithological effect is distributed along the vertical axis (SQp), while different fluid is distributed along the horizontal axis (SQs). Both lithological and fluid effects are separated orthogonally.

The crossplotting of SQp versus SQs can separate the lithology and pore fluid effects in 90 degrees. It shows that these attributes purely represent either the lithology effect or fluid effect and not both. This orthogonal separation between lithology and pore fluid is the same as what other methods such as the extended elastic impedance (EEI) method would have achieved. In the EEI method, the maximum separation is carried out by projecting the data in the fluid and lithology projection line by calculating the proper chi angle for the projection [10].

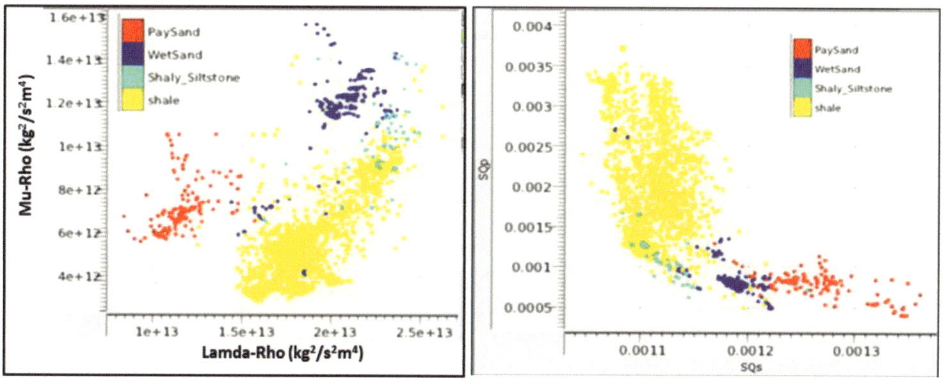

(a) (b)

Figure 7.
Lithological and fluid effect separation using crossplot method. (a) Mu-rho versus lambda-rho, (b) SQp versus SQs attribute. The lithology members consist of pay sand (gas sand), wet sand, shaly siltstone, and shale. Data taken from east Malaysian offshore.

Fortunately, in the SQp and SQp crossplotting, the projection line of lithology and fluid is constructed automatically as the orthogonal axis. This is one advantage when performing the interpretation using this attribute.

7. Application of SQp and SQs on fractured basement reservoir case

Previous tests were conducted on conventional reservoirs (clastic reservoir), with the objective being to test the sensitivity of SQp and SQs as indicators of lithology and fluid effect. To test the feasibility of SQp and SQs application on unconventional reservoir characterization, a well log data from fractured basement reservoir was taken from a field in the Malaysian Basin. In this well, the fracture is indicated by the image log. Image log is commonly practiced to detect the fracture of the borehole wall. However, this instrument is sensitive with diameter of borehole size. If there is a bad borehole condition of certain formation, the pad contact of the instrument with the borehole walls is not coupled properly. Hence, the fractures will not be effectively imaged. It is good to have another log as an alternative log that can be associated with the fractures which also can be derived from elastic properties. To fulfill the gap, the SQp and SQs attributes were tested by comparing it with the conventional brittleness average, fracture density log, and neutron porosity-density log to test the effectiveness of new attributes in terms of fracture density and hydrocarbon bearing identification in unconventional reservoir environment.

In this test, sample data was taken from depth 3125 to 3165 m of the fractured basement reservoir formation. The fracture density is compared to the brittleness average logs and SQP and SQS logs (**Figure 8**).

For this formation, the fracture density logs indicate the number of fracture on the formation. Brittleness average logs calculated from Poisson's ratio and Young's modulus was compared with fracture density logs. The results show that the brittleness average is consistent with the number of fracture density as shown in the fracture density logs. In the other side, the SQp log is also similar to the fracture density and brittleness average logs. High value of SQp is related with high fracture density and high brittleness average value. It shows that the SQp attribute also has

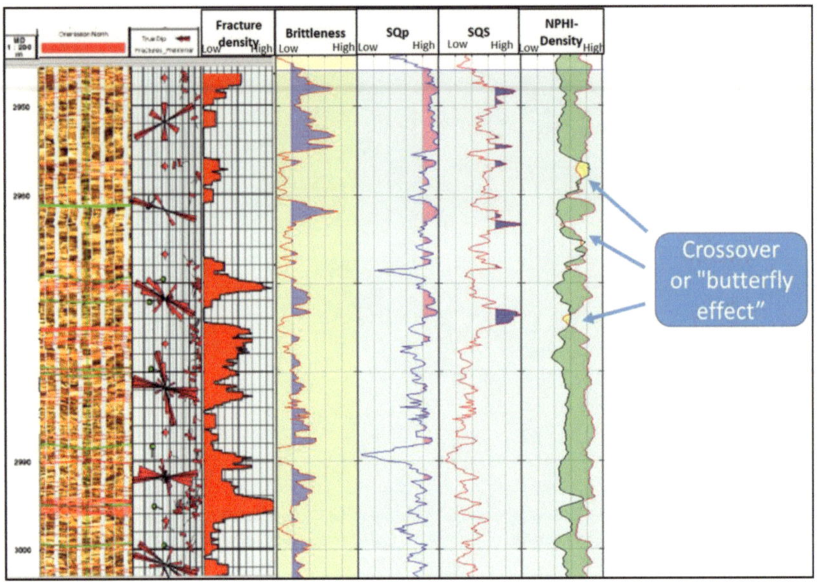

Figure 8.
Comparison between fracture density log, brittleness average log, SQP logs, SQS logs, and density-NPHI logs. Brittleness average looks consistent with fracture density log; SQp attribute is consistent with fracture density and SQs consistent with density-NPHI log.

potential to be used as a fracture density indicator or brittleness indicator in unconventional reservoir.

The example as shown in **Figure 8** was taken from an oil field-fractured basement reservoir. Conventional interpretation to identified oil column was conducted by interpreting the neutron porosity-density log together. Oil column will be identified as a crossover between neutron porosity and density log. This crossover is called also as "butterfly effect." The crossover of neutron porosity-density logs indicates the oil column. It is clear from **Figure 8** that the "butterfly effect" on neutron porosity-density log is associated with high value of SQs log. As mentioned in the previous test, the high value of SQs is indicating the hydrocarbon location. In this well, the SQs log is consistent with neutron porosity-density logs. It means that if the "butterfly effect" of neutron porosity-density log can be used to indicate hydrocarbon column in the fractured basement reservoir, the SQs attribute also can be used as indicator of hydrocarbon column for this unconventional reservoir environment.

From the test on this fractured basement reservoir well, both brittleness average, which is derived from Poison's ratio and Young's modulus and SQp attribute, have the same chance to be used as fracture density indicators, while the SQs attribute has the same potential as neutron porosity-density log in determining the location of hydrocarbon bearing. The difference is the SQs attribute can be extracted from seismic data, while the neutron porosity-density log can be analyzed on well log only. Hence, the use of SQs attributes can give us advantages to get the hydrocarbon distribution three-dimensionally.

The workflow to obtain the brittleness average and SQp and SQs attributes from seismic data is shown in **Figure 9**. A simultaneous inversion on pre-stack/partial-stack seismic data is needed to obtain P-wave (Vp), S-Wave (Vs), and density, which will be used to calculate the brittleness average and SQp and SQs attributes using Eqs. (5) and (10), respectively. An alternative method of obtaining the SQp and SQs attributes in a reflection domain can be approached using Eq. (14).

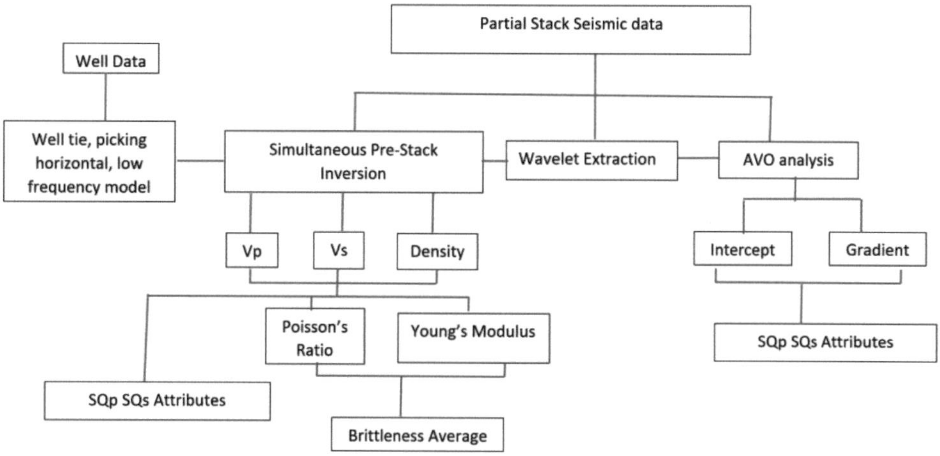

Figure 9.
Workflow for brittleness average and SQp and SQs attribute derivations from seismic data.

8. Conclusions

A common technique for brittleness estimation based on seismic elastic properties has been discussed. The existing method of brittleness average was tested and compared with new attributes (SQp) to indicate the existence of fracture density on a fractured basement reservoir environment in the Malaysian Basin. The results show that SQp attribute coincides with brittleness average and fracture density either on well log data or core data, while the SQs attribute is consistent with neutron porosity-density log which can be used to indicate the hydrocarbon column in the fractured basement reservoir.

Furthermore, the test of SQp and SQs attributes on conventional reservoir shows that those attributes are able to discriminate the lithological and fluid effects optimally. The lithology changes are indicated in the SQp log, which is similar to gamma ray log responses, while the fluid types are distinguished in the SQs log, which is similar to the resistivity log.

One of the advantages of using SQp and SQs attributes for reservoir characterization either in conventional or unconventional reservoir is that the attribute can be not only derived limited on well log domain but also applied three-dimensionally on seismic data. There are two options to get the SQp and SQs from seismic data: based on AVO analysis method and seismic inversion workflow. In the AVO method approach, the anomaly of hydrocarbon reservoir is indicated strongly in the SQp and SQs compared to the conventional AVO analysis using intercept and gradient crossplot method. The application of SQp and SQs attributes through inversion result also gives a strong indicator of lithology and fluid. Due to the similarity of SQp-SQs attributes with petrophysical properties, it is possible to use SQp and SQs attributes for petrophysical property prediction from elastic properties. This will become our future work.

Acknowledgements

We would like to thank UTP for supporting this research work and PETRONAS for providing the data.

Author details

Maman Hermana*, Deva Prasad Ghosh and Chow Weng Sum
PETRONAS University of Technology, Seri Iskandar, Malaysia

*Address all correspondence to: maman.hermana@utp.edu.my

References

[1] Perez R, Marfurt K. Calibration of brittleness to elastic rock properties via mineralogy logs in unconventional reservoirs. In: AAPG International Conference and Exhibition, vol. September; Cartagena, Colombia; 2013

[2] Jarvie DM, Hill RJ, Ruble TE, Pollastro RM. Unconventional shale-gas systems: The Mississippian Barnett shale of north-Central Texas as one model for thermogenic shale-gas assessment. American Association of Petroleum Geologists Bulletin. 2007;**91**(4):475-499

[3] Grieser WV, Bray JM. Identification of production potential in unconventional reservoirs. In: Production and Operations Symposium; 2007

[4] Leo CTAM. Exploration in the Gulf of Thailand in deltaic reservoirs, related to the Bongkot field. Geological Society London Special Publications. 1997; **126**(1):77-87

[5] Tjia AHD et al. In the quest of open fractures in the cystalline basement of the Malay and Penyu Basins. In: Petroleum Geology Conference and Exhibition. 2010; **2010**:124-126

[6] Mavko G, Mukerji T, Dvorkin JP. The Rock Physics Handbook. Calgary, Alberta: Society of Exploration Geophysicists; 2009

[7] Hermana M, Ghosh DP, Sum CW. Discriminating lithology and pore fill in hydrocarbon prediction from seismic elastic inversion using absorption attributes. The Leading Edge. 2017; **36**(11):902-909

[8] Castagna JP, Swan HW, Foster DJ. Framework for AVO gradient and intercept interpretation. Geophysics. 1998;**63**(3):948-956

[9] Gardner GHF, Gardner LW, Gregory R. Formation velocity and density—The diagnostic basics for stratigraphic. Geophysics. 1974;**39**(6):770-780

[10] Whitcombe DN, Connolly PA, Reagan RL, Redshaw TC. Extended elastic impedance for fluid and lithology prediction. In: Mavko G, Mukerji T, Dvorkin J, editors. The Rock Physics Handbook Tools for Seismic Analysis of Porous Media. 2nd edition. Cambridge University Press. 2009;**67**(1):63-67

Section 4

The Environment and Health and Safety

Chapter 7

Human Health Risks of Unconventional Oil and Gas Development Using Hydraulic Fracturing

Tanja Srebotnjak

Abstract

Advances in hydraulic fracturing technologies, and unconventional oil and gas (UOG) generally, spurred a boom in energy production in the United States. The rapid expansion of UOG has brought oil and gas production closer to homes, schools, and work places and thus increased potential human exposure to a range of chemicals, pollutants, and other health risks. Releases of such chemicals and pollutants occur throughout the full life cycle of UOG beginning with well-site preparation and continuing through hydraulic fracturing, well completion into production, well maintenance, and finally the plugging or abandoning of the well. While the risks to workers on UOG sites differ from those living, working or recreating nearby, both groups may be exposed to chemical and hazardous materials and injuries related to accidents and spills. This chapter characterizes the main occupational and public health risks throughout the life cycle of a hydraulically fractured well. Focusing on common practices in the United States, it identifies the main types of risks and pathways for human exposure. As a review, the chapter summarizes the peer-reviewed literature available to date, highlighting regulatory responses and identifying gaps in the current understanding of the risks involved in hydraulic fracturing.

Keywords: hydraulic fracturing, unconventional oil and gas development, health risk, air pollution, water pollution, occupational health risks, psychosocial health risks

1. Introduction

Advances in technologies that allow directional drilling coupled with high-volume hydraulic fracturing have made large unconventional oil and gas deposits accessible in the United States. The Energy Information Agency (EIA) estimates that in 2017 approximately 60% of U.S. dry natural gas production came from shale resources [1]. Similarly, oil production from tight oil formations rose from a negligible fraction in 2000 to 50% of total crude oil production in 2017 [2]. This growth has brought oil and gas production and related infrastructure closer to towns and communities in more than 20 states, with more than 15 active shale plays, and raised concerns about the risks to public health from chemical and nonchemical stressors associated with unconventional oil and gas production (UOG) [3]. This chapter provides a summary of these risks by taking a life cycle approach to characterizing the sources and types of health stressors and their likely exposure pathways. While UOG shares a number of processes with conventional oil and gas production, it differs in several important aspects, noticeably the use of directional (horizontal drilling) and large-volume hydraulic fracturing to stimulate the flow of natural gas or oil to the wellhead. These differences are particularly important as they pose additional, and to date not exhaustively regulated health, risks. There is also still greater uncertainty compared with conventional oil and gas production regarding the lack of information about the content of hydraulic fracturing fluids (HFF) and their health effects.

2. Life cycle risks of hydraulically fractured wells

As with any fossil fuel production, a hydraulically fractured ('fracked') well has the potential to release air and water pollutants, pose physical and public safety hazards, and contribute to psychosocial stressors for nearby residents and communities. The life cycle of a well consists of several phases shown in **Figure 1** [4].

Each life cycle phase generates emissions, effluents and waste that may pose health risks to workers and nearby communities. They are discussed in this chapter according to their exposure pathway, e.g., via air or water, and by exposed population groups, e.g., oil and gas workers or nearby communities. It is noted that the likelihood of health impacts is generally a function of the hazardousness of the chemical and nonchemical stressor (i.e., the stimulus causing undesirable health

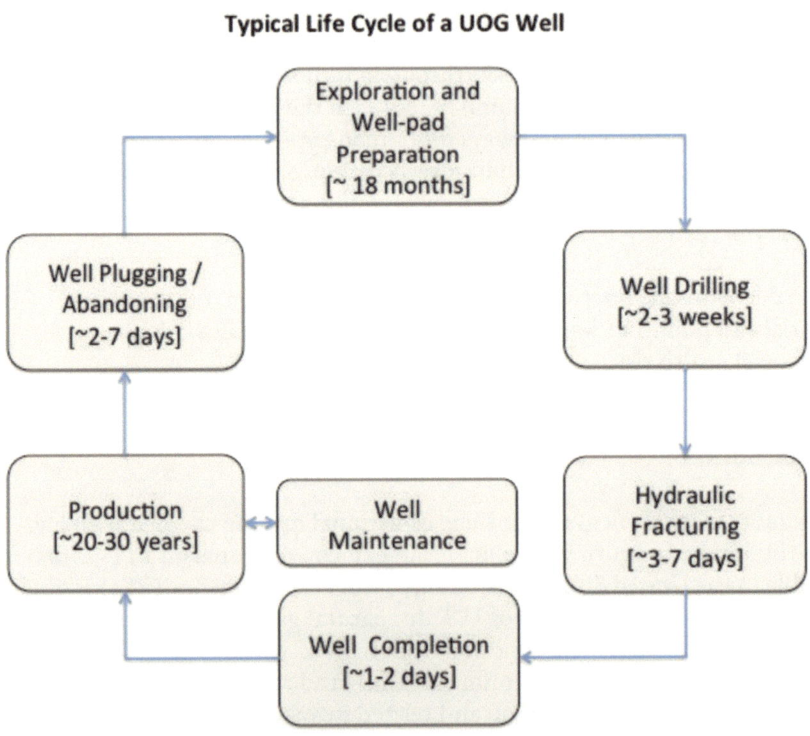

Figure 1.
The typical life cycle of an unconventional oil or gas well.

effects), the exposure duration and the pathway. The spatial reach of the stressor is also important and may range from the immediate well-pad area to local (up to 10 km), regional (up to 100 km) and local distances (farther than 100 km). Thus, the following sections are organized to describe human health risks according to pathway and spatial distance.

3. Occupational risks

The most significant types of occupational risks for UOG workers are accidents, malfunctions, and exposure to on-site air pollution. Accidents and malfunctions can bring workers in contact with hazardous and toxic materials via inhalation, dermal contact, or ingestion. They can also pose thermal radiation risks due to fires and explosions. Air pollution may be the result of accidents and malfunctions but is also a side effect of typical well-site activities such as oil and gas drilling, production, flaring, venting, storage of liquids and maintenance operations.

3.1 Accidents and unintentional releases

UOG wells are industrial sites with heavy and moving equipment, hazardous and toxic substances, and harsh environmental conditions. As a result, accidents and malfunctions (e.g., well blowouts, explosions, failure in well integrity such as sustained casing pressure and communication of the well with other, often orphaned wells) are the cause for most documented deaths and injuries for workers at unconventional well sites [5, 6].

Although it is difficult to obtain detailed information on worker-related injuries and fatalities for UOG, the oil and gas extraction industry in general has an occupational fatality rate that is 2.5-times that of the construction industry and 7-fold higher than the industry average [5, 6]. Fatalities are primarily caused by traffic-accidents (nearly a third of all confirmed fatalities), and smaller producers tend to have a higher mortality rate than larger and multi-national companies [7]. The traffic-related occupational risk to UOG workers is not surprising considering the substantial amount of material (e.g., water, HFF chemicals and additives, proppant), equipment (e.g., pipes, compressors, work-over equipment), and waste products (e.g., flowback and produced water, used drilling mud and drill cuttings) that need to be transported to and from the well site. Drilling and fracturing a well usually involves more than 1000 truck trips, often on narrow country roads not designed for such heavy use [8]. In contrast to UOG worker fatalities, the oil and gas industry has below-average injury rates, a fact that has been attributed to underreporting [6, 7, 9]. Although most accidents and fatalities occur among oil and gas workers, they also impact nearby communities. Truck accidents, well blowouts and explosions have caused injuries and fatalities among residents (see Section 4 for details).

3.2 Air pollution risks

The main sources of air pollution on UOG sites are [4]:

- Direct and fugitive emissions of methane and other hydrocarbons from well-heads and other production and transmission infrastructure on the well site (e.g., flowback and produced water holding tanks or evaporation ponds, valves, pipelines, processing equipment).

- Intentional venting and flaring of gas and hydrocarbon products.

- Diesel emissions from trucks, generators and diesel-powered equipment.
- Volatile organic compounds from drilling muds, HFF, flowback and produced water.

Workers may suffer from acute exposure to hazardous and toxic air pollutants such as hydrogen sulfide, benzene, formaldehyde and other volatile organic compounds [5]. Hydrogen sulfide arguably poses the greatest acute toxicity risk, causing irritation and central nervous system effects at concentrations as low as 100 ppm and death at around 1000 ppm [10]. Other risks arise from exposure to hydrocarbons, including aromatics such as benzene, ethylbenzene, toluene and the isomers of xylene (collectively referred to as BTEX). The health effects associated with BTEX include several types of leukemia, non-Hodgkin's lymphoma, anemia and other hematopoietic disorders, immunological effects, and reproductive and developmental effects [4, 11, 12]. While the health effects of BTEX are well documented and health-based regulatory exposure standards exist, other sources of exposure are less well characterized and not regulated. These include chemicals in HFF and volatilized components in drilling muds. A sizeable fraction of compounds used in HFF do not have Chemical Abstract Service (CAS) identifiers [13].

In addition, workers may suffer chronic exposure to stressors such as crystalline silica, which is the main proppant used in hydraulic fracturing to hold open rock fractures and ease the flow of oil and gas to the surface. Prolonged inhalation of silica can cause silicosis and lung cancer, and it is also associated with chronic obstructive pulmonary disease, kidney disease and autoimmune diseases [14]. OSHA has issued a health alert for workers concerning exposure to silica during hydraulic fracturing [15]. Esswein et al. reports a study by the National Institute for Occupational Safety and Health (NIOSH), which collected and analyzed 111 samples of personal breathing zone data for respirable crystalline silica exposure at 11 UOG sites in five states (Colorado, Texas, North Dakota, Arkansas, and Pennsylvania) [16]. They found that 93% of samples exceeded the threshold limit value (TLV) of the American Conference of Industrial Hygienists of 0.025 mg/m^3, 76% exceeded the Recommended Exposure Limit (REL) of the National Institute of Occupational Safety and Health (NIOSH) of 0.05 mg/m^3 and 51% were higher than the permissible exposure limit (PEL) by the Occupational Safety and Health Administration (OSHA) of 0.05 mg/m^3 averaged over an 8 hour-day. The differences in limits reflect the health protection goal of the respective institutions and the contexts and situations in which exposures are evaluated. Much of the silica sand (also known as 'frac sand') is mined in Wisconsin and Minnesota, thereby extending the occupational health risks to workers outside of the oil and gas industry and to regions where no hydraulic fracturing takes place [5]. Esswein et al. in a separate study also identified chemical exposure risks, including benzene, at six UOG sites in Colorado and Wyoming in 2013 and again found that wearable personal breathing zone monitors provided insufficient protection and were not always worn because of malfunctions [17].

3.3 Risks from soil contamination

Soil contamination from UOG operations can occur through surface spills of HFF, chemicals, drilling muds, and other compounds used during all life cycle phases of the well [18]. Health risks in these instances are largely limited to on-site workers and occur primarily through dermal contact. Workers may also carry contaminants indoors on their clothes and boots. Soil contamination has not yet been extensively studied in the UOG literature.

4. Non-occupational health risks

4.1 Accidents and unintentional releases

The heavy truck traffic associated with UOG, especially during the phases of well preparation, well drilling and hydraulic fracturing pose risks for vehicular accidents. In Bradford County, Pennsylvania, for example, the rise in truck traffic was concomitant with a rise in traffic accidents involving large trucks [8]. Similar statistics were observed in the Eagle Ford Shale in Texas. A study by Patterson et al. focusing on waste transport in the UOG sector in Pennsylvania found that UOG wells produced a median wastewater amount of 1294 m^3, requiring 122 heavy-truck trips for transportation off-site [19]. Throughout the full life cycle of a UOG well, and especially during the drilling and hydraulic fracturing stages, more than a thousand truck trips are required to transport water, chemicals, proppants, and equipment to and from the site. Since many well sites are now occupied by multiple wells, the health risks, such as air pollution from diesel engines and traffic accidents, increase even further.

Throughout most of the well's life cycle residents are also at risk of accidents due to malfunctions such as well blowouts, explosions, fire, spills, and leaks. These may release hazardous chemicals into the air and pose thermal radiation risks. Extreme weather events put oil and gas sites and associated infrastructure at risk. For example, holding and evaporation ponds for flowback and produced water overflowed in Colorado during the 2013 floods and released chemical and hydrocarbon laced liquids across the landscape and into nearby surface waters [20].

4.2 Air pollution risks

The around-the-clock operations of UOG production sites mean that people and communities in the vicinity may experience a continuous, albeit variable, exposure to airborne pollutants. In addition to infrequent but acute symptoms, they may thus suffer effects from cumulative exposure. The drilling, fracking and operation of UOG wells releases VOCs, from valves, pipes, condensate tanks, flowback and produced water tanks, and other infrastructure. Well maintenance operations such as offloading, additional fracking stages, etc. are often episodes of high air emissions of hydrocarbons, especially for natural gas wells. Residents have complained about odors and health symptoms such as headaches, nose bleeds, skin irritation, chronic fatigue, and neurological effects. A number of observational studies has shown associations between the occurrence of health symptoms and distance to the well, well density, and temporal coincidence with well-site activities [21–24]. Well completions, condensate storage tanks and compressors have been shown to release VOCs, including C2–C8 alkanes, aromatic hydrocarbons, methyl mercaptan, and carbon disulfide [4]. Also process-related is a study that found elevated concentrations of benzene, several aliphatic hydrocarbons in samples taken 130–500 feet from five well pads in Colorado during high-emission periods of uncontrolled flowback [4, 25]. The increased truck traffic also degrades local air quality through diesel exhaust, nitrogen oxides, dust, and other pollutants associated with diesel fuel combustion. Several studies of ambient air quality in densely populated areas with high UOG activity have shown that while the majority of wells produce emissions below regulatory standards and action levels, a few high-emitters can be responsible for the majority of emissions [26–28].

At the regional level, ozone, methane, benzene, and alkanes have been traced back to UOG production and installations, notably in Colorado's Front Range, the Denver-Julesburg Basin, the Niobrara Basin, the Uintah Basin, and the Upper Green

River Basin [29–32]. Winter ozone levels in some of these regions have reached levels (149 ppb) exceeding the worst days of day-time ozone levels in Los Angeles, one of the most ozone-polluted cities in America. Emission inventories showed that 98–99% of the VOCs and 57–61% of NO_x were attributable to unconventional oil and gas production [33]. Texas and Louisiana are also projecting increases in ground-level ozone concentrations of between 9 and 17 ppb above current concentrations for the low and high-emission scenarios, which may push some counties into non-attainment status of the federal ozone air quality standard (70 ppb).

Global effects of the growth in UOG arise from increases in methane emissions. Bottom-up and top-down studies have revealed higher methane levels in areas with UOG production, mostly natural gas shale plays, than under previous emission inventories released by the U.S. Environmental Protection Agency (EPA). According to the Intergovernmental Panel on Climate Change (IPCC) 5th Assessment Report (AR5), methane has a global warming potential that is 28 times that of carbon dioxide over a 100-year time horizon. Thus, while the transition of electric power generation from old, dirty coal-fired power plants to more efficient and cleaner natural gas plants is associated with regional air quality improvements, the climate benefits of UOG for shale gas remain somewhat disputed [34–37].

4.3 Water pollution risks

Most of the public attention surrounding hydraulic fracturing concerns the risks of surface and groundwater pollution, especially in the context of drinking water wells. There are several pathways for such pollution [4], including:

- Surface spills on-site and during transportation involving HFF, liquid drilling mud, and chemicals.
- Well casing leaks.
- Migration of gases and liquids through fractured rock into groundwater aquifers and to the surface.
- Leaks from abandoned wells.
- Wastewater discharges on-site or at wastewater treatment facilities.

A study by Gross for Colorado found that water pollution from surface spills is a relatively frequent occurrence: groundwater were impaired in 77 surface spills that were reported between July 2010–July 2011 and representing ~0.5% of active wells in densely drilled Wells County, Colorado [38]. Such impairment occurs primarily when spilled fluids percolate through the soil into shallow groundwater aquifers.

Vengosh et al. undertook a detailed study to understand which pathways were most likely for surface and subsurface migration. They distinguished between (i) the contamination of shallow aquifers with fugitive hydrocarbon gases (stray gas contamination, (ii) contamination of surface water and shallow groundwater from spills, leaks, and/or the disposal of inadequately treated shale gas wastewater, and (iii) the accumulation of toxic and radioactive elements in soil or stream sediments near disposal spill sites [3]. Using published data and studies from across the U.S. the results indicate that there is evidence for stray gas contamination, surface water impacts, and accumulation of radium isotopes at some disposal and spill sites. A critical issue in conclusively attributing the pollution of drinking water wells or other water sources to UOG operations is the lack of baseline data, i.e., data on water quality before UOG commenced. In particular, methane, heavy metals, and radioactive compounds may have been in the water long before the arrival of unconventional oil and gas production as a result of the aquifers' geology or due to other man-made activities. In order to better understand and attribute the sources of water contamination, states such as California now require water quality monitoring before and after unconventional well drilling and stimulation activities. Overall, the evidence for methane contamination of groundwater and drinking water wells from hydraulic fracturing remains controversial in many cases [39–41].

4.4 Water resource depletion

While the debate and study of how water quality may be impacted by UOG activities continues, there is clear evidence that water abstraction for hydraulic fracturing in water-scarce areas can lead to increased competition and shortages. A report by Ceres, a sustainability non-profit organization formerly known as the Coalition for Environmentally Responsible Economies, examined the relationship between water use by UOG and other sectors in water-stressed regions and found that water resources are negatively impacted [42]. Nationwide, Ceres looked at nearly 110,000 UOG wells and estimated that 57% of oil and gas wells hydraulically fractured between 2011 and 2015 were in water-scarce regions and where water is a subject of competition among farmers, towns and cities, and the oil and gas industry. Overall, fracking-related water use during the 5-year study period totaled 358 billion gallons. Put into perspective, this amount of water is consumed by approximately 200 mid-sized U.S. cities. States with significant oil and gas production that are particularly impacted by the threat of increasing water competition are Texas, Colorado, and California. These states are also home of some of the leading shale plays, including the Eagle Ford and the Midland Play (part of the Permian Basin) in Texas. Other plays characterized by high water use are the Marcellus Shale and the Niobrara in Colorado, Wyoming, and Nebraska. The county with the highest number of UOG wells (~7000 wells) and water use for hydraulic fracturing (>16 billion gallons) is Weld County, CO [42].

4.5 Waste management and disposal risks

Wastewater is the largest waste stream in oil and gas production. It consists primarily of produced water, which is for the most part brine mixed with hydrocarbons and suspended solids. Produced water is distinct from flowback water, which consists primarily of HFF and is generated for the first few days after hydraulic fracturing operations. Produced water may contain chemicals and additives used in drilling mud, methane, petroleum condensate, heavy metals, naturally occurring radioactive materials (NORM). Typically, flowback and produced water are temporarily stored in on-site pits (also called evaporation sumps) or tanks prior to disposal or reuse/recycling. These pits pose air and water pollution risks, from the release of volatile compounds into the air (evaporation is the purpose of some pits and may be supported by aerators) and the use of unlined pits. Flooding can also lead to pits overflowing and dispersing their hazardous contents across the landscape and potentially contaminating groundwater and nearby surface streams. On-site spills due to broken pipes or deterioration of the exterior walls of pits can also lead to localized soil and water contamination. California's oil-rich Central Valley has a legacy of unlined pits and at least one of the sites is known to have a sub-surface pollution plume that is threatening the Kern River [43].

The majority of produced water is disposed of through deep-injection wells (class II wells according to the UIC program by EPA). In Pennsylvania, produced water was initially send to publicly owned treatment works, but the treatment processes were not adequate to handle the high TDS and chemical-contaminated water and the state prohibited the practice. If the injection wells reach aquifers that may potentially be used as a source of water for drinking or other purposes, the practice may threaten the water supply in water-stressed regions. This is the case in California, where hundreds of injection wells were found to be in potential violation of the Safe Drinking Water Act [44].

4.6 Socioeconomic and psychosocial risks

In addition to the potential health benefits arising from the replacement of coal-fired power plants with plants using shale gas, the development of UOG can generate local and regional economic benefits that can improve the overall health of the population. Additional jobs and UPG producer taxes can stimulate the economy and lead to greater public investments in education and healthcare. The estimates vary but UOG related employment in the U.S. might be in the order of 1.7 million people with further growth projected [45]. However, negative health effects of UOG expansion can occur from "boomtown" effects, i.e., the extractive-resource driven rapid expansion of local economies, which is followed by equally fast declines when resource prices decline or other market forces throttle production. These effects have been well-documented in the 1970s and 1980s and they tend to hit the most vulnerable members of communities first and hardest [4]. Psychosocial studies have also documented the tensions that can rise in communities split into supporters and opponents of UOG, by concerns of the local residents about known and unknown side-effects of UOG production such as air and water contamination, and concern over social disruptions due to the sudden influx of mostly male workers from other parts of the country. Residents surveyed in rural parts of Pennsylvania, where Marcellus Shale development has grown rapidly, have mentioned feelings of loss concerning their old way of living, the degradation of pristine environments, and their sense of place. Psychosocial stress can manifest itself in a variety of symptoms that are difficult to diagnose. They can be exacerbated by a lack of trust in the UOG producers and local and state government regarding the safe development of these unconventional resources. The Geisinger Health System in Pennsylvania is undertaking a series of coordinated studies of the population it serves regarding self-reported symptoms [46].

5. Health research needs

Considering the diverse range of potential health risks emanating from UOG operations and the role that local factors such as regulations, geology, climate, proximity to population centers, etc. play, there are a number of open research questions that should be addressed. Arguably the most pressing issue is the lack of information about UOG activities followed by the need for toxicological and epidemiological studies.

In particular, the major information gaps and uncertainties regarding our understanding of the health risks of UOG development impact the ability of regulators, healthcare professionals, communities, and individuals to take appropriate measures to protect against them, to inform others, and to work with the industry to mitigate the negative effects of UOG. The most important issues to be addressed from a research and data development perspective are shown in **Table 1**.

These gaps and uncertainties should be systematically addressed in future studies, which require improved cooperation between UOG producers, federal,

Occupational health and safety	Public health
Study the occupational health and safety risks of UOG with respect to the use and regulation of personal protective equipment and the activities that put oil and gas workers at most risk.	Collect and release timely and complete information about the composition of hydraulic fracturing fluids, in particular, addressing the use of trade secret protections and the accessibility of information by emergency responders, public health officials, oil and gas regulators, the scientific community, and the general public.
Assess the differences of UOG activities compared with conventional oil and gas development and determine targeted and effective occupational health protections for them.	Continue efforts to close the knowledge gap on the health risks of UOG through toxicological and exposure-effect studies of HFF constituents.
Fix the incomplete reporting of occupational health and safety incidents, especially with respect to injuries in order to reduce underreporting in state and federal statistics.	Conduct longer-term epidemiological studies to improve the scientific understanding of the associations and causalities between exposure to UOG-related hazards and reported health symptoms. These include the systematic description and assessment of exposure pathways and severity and their duration as well as developing an improved understanding of the effects of multiple well sites with regard to cumulative and aggregate exposures.
	Invest in community-based studies on the psychosocial stresses and associated health outcomes resulting from the expansion of UOG activities into rural communities.
	Develop databases and systematic guidelines for air and water quality monitoring before and during UOG activities with the goal to improve source-attribution in cases of deteriorated air and/or water quality.
	Develop tracers and other solutions to better identify and attribute the causes of drinking water well contamination in the context of UOG activities.

Table 1.
List of proposed research and development activities needed to fully understand and mitigate the risks of UOG on worker and public health.

state and local government, and community health and environmental advocacy groups. FracFocus, an industry-sponsored database providing information on hydraulically fractured wells and the HFF used, is a step towards addressing this information gap, but it is voluntary, incomplete, and lacks some important functionality.

Occupational health risks would benefit from better surveillance and reporting, especially of injuries. Focus should be on monitoring exposure to benzene, toluene, silica, aliphatic hydrocarbons, diesel exhaust, HFF chemicals, hydrogen sulfide, NORM, and traffic related exposures [4]. In addition, studies on both chronic and acute exposures are relevant in the occupational health context.

The proposed before-and-after monitoring of water and air quality, in the context of planned UOG development, could provide a stronger foundation to accurately and conclusively determine if contamination events occurred and what their source was. The variable and locally specific context of UOG development calls for studies that assess the magnitude and duration of human exposure to stressors during the various life cycle phases of UOG wells. For example, HFF mixture, geology, type of unconventional resource, and environmental factors all influence the potential for exposure and resulting health effects. In addition, the dense clustering of wells typical for UOG development creates the risk of aggregate effects that need to be further assessed.

With regard to psychosocial and community health impacts, current knowledge could be enhanced through greater involvement of community organizations as a source of information and for building trust between community members on the one hand and scientists, public health officials, and regulators on the other. These community organizations can furthermore serve as a bridge for continued outreach, education, and data collection after studies have been completed. UOG producers could rebuild trust by actively engaging with the community in the planning processes of new UOG development, providing fact-based information about the development, and supporting community activities aimed at identifying and reducing sources of stress. Public-private partnerships, such as the Health Effects Institute, have been able to bridge the trust-deficit and can serve as a model for working to solve the often-contentious health issues [4].

6. Conclusions

The use of large-volume, horizontal hydraulic fracturing has expanded across the U.S. and inspired talk of American energy independence and a renaissance of manufacturing. At the same time hydraulic fracturing has also become a lightning rod in public debates that pitches neighbors against each other and prompted calls for moratoria and greater scientific scrutiny from environmental groups and community health advocates. This chapter is an attempt to summarize the main sources of environmental pollution and health risks that arise during the lifespan of a hydraulically fractured well. It is a reminder that the reader that unconventional oil and gas production is an industrial activity that is noisy, dirty, and that generates substantial amounts of waste. Some of these side effects occur primarily on the well pad and in its immediate vicinity, where they pose risks to workers and residents. Others manifest themselves regionally and even globally and thus add to the pollution burden of people and communities who are far away from oil and gas production. The regulatory environment in which oil and gas development takes place usually creates obstacles for people to receive information and seek redress for pollution and health effects they might experience. Indeed, the burden of proof of causality between unconventional oil and gas operations as the source of the impacts is often on the individual or community and requires a level of scientific knowledge and information that is beyond their capacity. This is where regulators, public health officials, and the scientific community need to focus and together with the oil and gas industry develop mechanisms for greater transparency, meaningful data collection, and targeted epidemiological and toxicological studies. Unconventional oil and gas development is projected to continue its growth path and will remain a part of life in many rural and also urban communities across the U.S. In order to facilitate a co-existence that is based on trust, prioritizes safety over profits, and invests in local communities, the discussed health risks need to be addressed comprehensively and form the evidentiary basis for regulatory action.

Conflict of interest

I declare that I have no conflict of interest.

Author details

Tanja Srebotnjak
Harvey Mudd College, Claremont, California, USA

*Address all correspondence to: tsrebotnjak@g.hmc.edu

References

[1] EIA. 2018. Available from: https://www.eia.gov/tools/faqs/faq.php?id=907&t=8 [Accessed: Sep 3, 2018]

[2] EIA. 2018. Available from: https://www.eia.gov/tools/faqs/faq.php?id=847&t=6 [Accessed: Sep 3, 2018]

[3] Vengosh A, Jackson RB, Warner N, Darrah TH, Kondash A. A critical review of the risks to water resources from unconventional shale gas development and hydraulic fracturing in the United States. Environmental Science & Technology. 2014;**48**(15):8334-8348

[4] Adgate JL, Goldstein BD, McKenzie LM. Potential public health hazards, exposures and health effects from unconventional natural gas development. Environmental Science & Technology. 2014;**48**(15):8307-8320

[5] Witter RZ, Tenney L, Clark S, Newman LS. Occupational exposures in the oil and gas extraction industry: State of the science and research recommendations. American Journal of Industrial Medicine. 2014;**57**(7): 847-856

[6] Bureau of Labor Statistics. Census of Fatal Occupational Injuries. Washington, DC; 2009. Available from: http://www.bls.gov/iif/oshcfoi1.htm

[7] Retzer KD, Hill RD, Pratt SG. Motor vehicle fatalities among oil and gas extraction workers. Accident Analysis & Prevention. 2013;**51**:168-174

[8] NY Department of Environmental Conservation. Revised Draft SGEIS on the Oil, Gas, and Solution Mining Regulatory Program: Well Permit Issuance for Horizontal Drilling and High Volume Hydraulic Fracturing in the Marcellus Shale and Other Low Permeability Gas Reservoirs. New York State; 2011. Available from: http://www.dec.ny.gov/data/dmn/rdsgeisfull0911.pdf

[9] Mendeloff J, Burns R. States with low non-fatal injury rates have high fatality rates and vice-versa. American Journal of Industrial Medicine. 2013;**56**(5):509-519

[10] ATSDR. Toxicological Profile of Toluene. 2018. Available from: https://www.atsdr.cdc.gov/toxprofiles/TP.asp?id=161&tid=29 [Accessed: Sep 12, 2018]

[11] ATSDR. Toxicological Profile of Ethylbenzene. 2010. Available from: https://www.atsdr.cdc.gov/toxprofiles/TP.asp?id=383&tid=66 [Accessed: Sep 12, 2018]

[12] ATSDR. Toxicological Profile of Hydrogen Sulfide/Carbonyl Sulfide. 2018. Available from: https://www.atsdr.cdc.gov/toxprofiles/tp.asp?id=389&tid=67 [Accessed: Sep 12, 2018]

[13] Colborn T, Kwiatkowski C, Schultz K, Bachran M. Natural gas operations from a public health perspective. Human and Ecological Risk Assessment: An International Journal. 2011;**17**(5):1039-1056

[14] ATSDR. Toxicological Profile of Silica. 2017. Available from: https://www.atsdr.cdc.gov/toxprofiles/TP.asp?id=1483&tid=290 [Accessed: Sep 12, 2018]

[15] OSHA. 2018. Available from: https://www.osha.gov/dts/hazardalerts/hydraulic_frac_hazard_alert.html [Accessed: Aug 29, 2018]

[16] Esswein EJ, Breitenstein M, Snawder J, Kiefer M, Sieber WK. Occupational exposures to respirable crystalline silica during hydraulic fracturing. Journal

of Occupational and Environmental Hygiene. 2013;**10**(7):347-356

[17] Esswein EJ, Snawder J, King B, Breitenstein M, Alexander-Scott M, Kiefer M. Evaluation of some potential chemical exposure risks during flowback operations in unconventional oil and gas extraction: Preliminary results. Journal of Occupational and Environmental Hygiene. 2014;**11**(10):D174-D184

[18] Finkel ML, Law A. The rush to drill for natural gas: A public health cautionary tale. American Journal of Public Health. 2011;**101**(5):784-785

[19] Patterson LA, Maloney KO. Transport of hydraulic fracturing waste from Pennsylvania wells: A county-level analysis of road use and associated road repair costs. Journal of Environmental Management. 2016;**181**:353-362

[20] CO Oil and Gas Conservation Commission. COGCC Flood Information. 2013. Available from: http://cogcc.state.co.us/Announcements/Hot_Topics/Flood2013/Flodd.htm

[21] McKenzie LM, Witter RZ, Newman LS, Adgate JL. Human health risk assessment of air emissions from development of unconventional natural gas resources. Science of the Total Environment. 2012;**424**:79-87

[22] Ferrar KJ, Kriesky J, Christen CL, Marshall LP, Malone SL, Sharma RK, et al. Assessment and longitudinal analysis of health impacts and stressors perceived to result from unconventional shale gas development in the Marcellus Shale region. International Journal of Occupational and Environmental Health. 2013;**19**(2):104-112

[23] Steinzor N, Subra W, Sumi L. Investigating links between shale gas development and health impacts through a community survey project in Pennsylvania. New solutions: A Journal of Environmental and Occupational Health Policy. 2013;**23**(1):55-83

[24] Rabinowitz PM, Slizovskiy IB, Lamers V, Trufan SJ, Holford TR, Dziura JD, et al. Proximity to natural gas wells and reported health status: Results of a household survey in Washington County, Pennsylvania. Environmental Health Perspectives. 2014;**123**(1):21-26

[25] Colborn T, Schultz K, Herrick L, Kwiatkowski C. An exploratory study of air quality near natural gas operations. Human and Ecological Risk Assessment: An International Journal. 2014;**20**(1):86-105

[26] Pring MO, Wilhelmi J. Fort Worth Natural Gas Air Quality Study. 2017. Available from: https://www3.epa.gov/ttnchie1/conference/ei20/session6/mpring.pdf [Accessed: Sep 12, 2018]

[27] ERG. City of Fort Worth Natural Gas Air Quality Study. Final Report. Eastern Resources Group, Inc. (ERG); 2011. Available from: http://fortworthtexas.gov/uploadedFiles/Gas_Wells/AirQualityStudy_final.pdf [Accessed: Sep 12, 2018]

[28] Zielinska B, Campbell D, Samburova V. Impact of emissions from natural gas production facilities on ambient air quality in the Barnett Shale area: A pilot study. Journal of the Air & Waste Management Association. 2014;**64**(12):1369-1383

[29] Mallula S. Effects of non-conventional gas and oil production activities on local and regional fine particles and ground-level ozone [doctoral dissertation]. The University of North Dakota; 2017

[30] Ahmadov R, McKeen S, Trainer M, Banta R, Brewer A, Brown S, et al. Understanding high wintertime ozone pollution events in an oil-and natural gas-producing region of the western

US. Atmospheric Chemistry and Physics. 2015;**15**(1):411-429

[31] Helmig D, Thompson CR, Evans J, Boylan P, Hueber J, Park JH. Highly elevated atmospheric levels of volatile organic compounds in the Uintah Basin, Utah. Environmental Science & Technology. 2014;**48**(9):4707-4715

[32] Pétron G, Karion A, Sweeney C, Miller BR, Montzka SA, Frost GJ, et al. A new look at methane and nonmethane hydrocarbon emissions from oil and natural gas operations in the Colorado Denver-Julesburg Basin. Journal of Geophysical Research: Atmospheres. 2014;**119**(11):6836-6852

[33] Utah Dept. of Environmental Quality. 2012 Uintah Basin Winter Ozone & Air Study. Final Report. 2012. Available from: http://rd.usu.edu/files/uploads/ubos_2011-12_final_report.pdf [Accessed: Sep 12, 2018]

[34] Howarth RW, Santoro R, Ingraffea A. Methane and the greenhouse-gas footprint of natural gas from shale formations. Climatic Change. 2011;**106**(4):679

[35] Jiang M, Griffin WM, Hendrickson C, Jaramillo P, Van Briesen J, Venkatesh A. Life cycle greenhouse gas emissions of Marcellus Shale gas. Environmental Research Letters. 2011;**6**(3):034014

[36] Burnham A, Han J, Clark CE, Wang M, Dunn JB, Palou-Rivera I. Life-cycle greenhouse gas emissions of shale gas, natural gas, coal, and petroleum. Environmental Science & Technology. 2011;**46**(2):619-627

[37] O'Sullivan F, Paltsev S. Shale gas production: Potential versus actual greenhouse gas emissions. Environmental Research Letters. 2012;**7**(4):044030

[38] Gross SA, Avens HJ, Banducci AM, Sahmel J, Panko JM, Tvermoes BE. Analysis of BTEX groundwater concentrations from surface spills associated with hydraulic fracturing operations. Journal of the Air & Waste Management Association. 2013;**63**(4):424-432

[39] Osborn SG, Vengosh A, Warner NR, Jackson RB. Methane contamination of drinking water accompanying gas-well drilling and hydraulic fracturing. Proceedings of the National Academy of Sciences. 2011;**108**(20):8172-8176

[40] Davies RJ. Methane contamination of drinking water caused by hydraulic fracturing remains unproven. Proceedings of the National Academy of Sciences. 2011;108(43). https://doi.org/10.1073/pnas.1113299108

[41] Jackson RB, Vengosh A, Darrah TH, Warner NR, Down A, Poreda RJ, et al. Increased stray gas abundance in a subset of drinking water wells near Marcellus Shale gas extraction. Proceedings of the National Academy of Sciences. 2013;**110**(28):11250-11255

[42] Ceres. New Data: Water Use in Hydraulic Fracturing a Key Risk in Water Stressed Regions in Colorado and Texas. 2016. Available from: https://www.ceres.org/news-center/press-releases/new-data-water-use-hydraulic-fracturing-key-risk-water-stressed-regions [Accessed: Sep 12, 2018]

[43] SF Chronicle. Pits of Drilling Waste threaten Water, Air Quality, Report says. 2016. Available from: https://www.sfchronicle.com/science/article/Pits-of-drilling-waste-threaten-water-air-6873892.php [Accessed: Sep 12, 2018]

[44] EPA. EPA's Oversight of California's UIC Program. 2018. Available from: https://www.epa.gov/uic/epas-oversight-californias-underground-injection-control-uic-program [Accessed: Sep 12, 2018]

[45] IHS. The Economic and Employment Contribution of Unconventional Gas Development in State Economies. Washington, DC; 2012. Available from: http://marcelluscoalition.org/wp-content/uploads/2012/06/State_Unconv_Gas_Economic_Contribution_Main.pdf [Accessed: Sep 12, 2018]

[46] Geisinger Caring. Unconventional Natural Gas Development. 2018. Available from: https://www.geisinger.edu/research/departments-and-centers/environmental-health-institute/unconventional-natural-gas-development [Accessed: Sep 12, 2018]